Archer Fish Biology

Simon Kumar Das

Faculty of Science and Technology
National University of Malaysia
Bangi, Selangor, Malaysia

CRC Press
Taylor & Francis Group
Boca Raton London New York

CRC Press is an imprint of the
Taylor & Francis Group, an **informa** business

Cover image provided by Shelby Temple, PhD. (Bristol, United Kingdom).

First edition published 2024
by CRC Press
2385 NW Executive Center Drive, Suite 320, Boca Raton FL 33431

and by CRC Press
4 Park Square, Milton Park, Abingdon, Oxon, OX14 4RN

Library of Congress Cataloging-in-Publication Data (applied for)

ISBN: 978-0-367-46237-6 (hbk)
ISBN: 978-1-003-02767-6 (ebk)

DOI: 10.1201/9781003027676

Typeset in Palatino
by Innovative Processors

Preface

This book is a compilation of chapters that discuss in depth knowledge about the basic biology of archer fishes. The topics included in it are of utmost significance and bound to provide incredible insights about the basic biology of two congeneric archer fishes, especially their age and growth, food and feeding, feeding mechanisms, digestion physiology and reproductive biology to readers. The level of detail provided in this work enables readers to effectively utilise primary research papers pertaining to these intriguing fish species. Furthermore, it has the potential to serve as an exemplary textbook for students specialising in fish biology.

A foreword of all chapters of the book is provided below:

Chapter 1 – This chapter is an overview of the subject matter incorporating all the major aspects of the biology of two congeneric archer fishes including their morphology, identification techniques, feeding physiology, and reproduction, **Chapter 2** – The topics discussed in this chapter are of great importance to broaden the existing knowledge on the identification techniques of archer fish species. **Chapter 3** – The information on population age and growth, and condition factors are discussed in this chapter. **Chapter 4** – This chapter describes the phenomenal feeding techniques of two congeneric archer fishes including their spitting success and accuracy according to their size classes. Besides that, this chapter discusses the diet and identifies the trophic level of archer fishes from the wild samples. **Chapter 5** – Knowledge of fish gastric evacuation rate is a necessary component for both field and laboratory studies concerned with fish feeding rates, energy budgets, and the trophic dynamics of aquatic systems. This can be achieved by tracing the movement of food item in the gut passage using X-radiography and serial slaughter techniques. The present chapter also facilitates the modelling approach to determine the gastric emptying rate in archer fish. The absorption of macronutrients passing through the alimentary tract of *T. jaculatrix* conditioned to feed on whole mealworm were also studied. **Chapter 6** – Overall reproductive biology of *T. chatareus* and *T. jaculatrix* are discussed in this chapter.

Acknowledgements

My heartiest thanks to Ministry of Science, Technology and Innovation, Malaysia (MOSTI) and Universiti Kebangsaan Malaysia for being able to fully sponsor this work through UKM Science fund (No. 04-01-02-SF0124) and the fellowship of skim University Kebangsaan Malaysia (No. UKM-OUP-FST-2008/2009/2010).

I am sincerely indebted and thankful to Professor Dr. Mazlan Abd. Ghaffar, whose intellectual encouragement, guidance, and invaluable academic and personal assistance made this work possible.

I am extremely grateful to Dr. Shelby Temple and Dr. Yosni Bakar, both of whom read and commented extensively on the entire manuscript, and cared enough about it to be remorselessly critical.

Thanks also to Prof. Brian C. Small, Dr. N.J. Brown-Peterson, Prof. Gerald R. Allen and Professor Malcolm Jobling for the valuable discussion and for expertly commenting on sections of the text.

Heartiest thanks to Mr. Azarindra and Mr. Husdy, who over a number of years have helped me, collect fish from the mangrove areas. Thanks for their friendship, for sharing their knowledge of the sea and for many enjoyable fishing days.

Thanks to Mr. Bakri in Chemistry lab, Mrs. Sawiyah in Food Science lab and Miss Salvana in Scanning Microscopy lab UKM, for all the assistance they extended to me regarding various aspects of my lab works.

I extend my heartfelt thanks to the dedicated editorial team of CRC Press, Taylor and Francis group for their unwavering patience and support during the publication of this book.

A note of appreciation to Dr. Siti Zahrah Abdullah and Oo Mooi Gaik, National fish health research centre, Penang, Malaysia for their considerable cooperation and assistance during the histology works. Thanks to my students Noorashikin for her help formatting this thesis.

And finally, on a much more personal note, I would like to express my immense gratitude to my parents, elder brother, and sister, my wife, and daughters, to whom this piece of work is dedicated. I could not thank them enough.

Simon Kumar Das

List of Tables

List of Figures

List of Abbreviations

ADC	Apparent digestibility coefficient
AFR	Anal fin ray
AM	Ante meridiem: before noon
Ant. int.	Anterior intestine
AS	Anal spine
BD	Body depth
BW	Body width
ca	Cortical alveoli
ca	Circa: about, approximately
CFL	Caudal fin length
CFR	Caudal fin ray
cm	Centimeter
cn	Chromatin nucleolar oocyte
DCP	Depth of caudal peduncle
DFA	Discriminant function analysis
DFR	Dorsal fin ray
DM	Dry mass
e.g.	Exempli gratia: for example
EB	Empty body weight
ED	Eye diameter

et al.	Et alia: and others
etc.	Et cetera: and so forth
EW	Eviscerated weight
F	Fecundity
fo	Frequency of occurrence
g	Gram
GE	Gastric emptying/evacuation
GERs	Gastric emptying rates
GET	Gastric emptying time
GSI	Gonadosomatic index
GW	Gonad weight
ha.	Hectare

List of Symbols

≤	Less than or equal to
≥	Greater than or equal to
$BaSO_4$	Barium sulphate
Cs	Caesium
^{14}C	Carbon-14
^{51}Cr	Chromium-51
^{137}Ce	Cerium-137
F-Ni	Ferro-nickel
^{131}I	Iodine-131
^{15}N	Nitrogen-14
^{99m}Tc	Technetium-99m
HCl	Hydrochloric acid
K	Fulton condition factor, growth coefficient
K_n	Relative condition factor
L_∞	Asymptotic length
ϕ'	Phi prime
°C	Degree Celsius
r	Correlation coefficient
r^2	Coefficient of determination
t_o	Hypothetical age

κ	Kappa
λ, Λ	Lambda
χ	Chi
%	Percentage
ρ	Gastric emptying rate
α	Gastric emptying coefficient
ξ	Random error term
μ	Mu

Contents

Archer Fish

Introduction

Fish are by far the most diverse of all vertebrate groups, with the number of their recognised species exceeding 25,000 and continuing to increase as new species are described (Bone & Marshall 1982; Eschmeyer & Froese 2003; O'Dor 2003). They can be found in diverse aquatic environments, with certain species spending a significant portion of their time outside of water. The diversity of fishes is highest in the tropics, where, in marine waters, it is greatest in the shallow waters of continental shelves (Nelson 1994).

Mangroves, which are often found in tropical estuaries, are highly productive ecosystems (Blaber 2000; Robertson et al. 1992). Consequently, mangroves are an important habitat for fish, providing both an abundance of food and shelter from predation (Cocheret de la Morinière et al. 2004; Laegdsgaard & Johnson 1995). The structural complexity and heterogeneity of mangrove stands, in combination with the range of potential food they contain, accounts for the diversity of the fish assemblage found in this type of habitat (Laegdsgaard & Johnson 2001; Lugendo et al. 2006; Rönnbäck et al. 1999; Sasekumar et al. 1992). Archer fish inhabit mangrove habitat and they have acquired the remarkable skill to hunt for prey outside of the water, by shooting a jet of water with pinpoint accuracy to knock off stationary insects in the mangrove foliage above (Allen 2004; Gill 1909). Perhaps this unique feeding behaviour was important in ensuring continuous survival and resilience of these fishes, since they were first described nearly 250 years ago (Schlosser 1764). Essentially very few life history works have been conducted on them. *Toxotes chatareus* (Hamilton) and *Toxotes jaculatrix* (Pallas) are the two species of archer fishes found in the coastal waters of Malaysia and other areas from Sri Lanka and India to New Guinea and northern Australia (Allen 1978, 2001; Day 1971). The morphological features of archer fishes are scarce and restricted to structures of the buccal cavity (Smith 1945), to the skeleton (Gill 1909), and to the adductor arcus palatini muscle (Milburn & Alexander 1976). However, the cranial structures of *Toxotes chatareus* are ornately described by Elshoud & Koomen (1985). Apart from the distribution ranges and morphological features, the reproductive biology of these fishes is hardly known. Pethiyagoda (1991) and Allen (2002) reported that *T. chatareus* females are highly fecund and release between 20,000 and 150,000 eggs. The taxonomic position of archer fishes is precisely described by Allen (2004). The tolerable salinity ranges (0-35 psu) of archer fishes are reported by Barletta et al. (2005).

The archer fishes *Toxotes chatareus* and *Toxotes jaculatrix* along with all members of the family Toxotidae, are well known for their amazing prey catching techniques and thus data are readily available (Bekoff & Dorr 1976; Dill 1977; Gill 1909; Lüling 1955, 1958, 1963, 1964, 1969; Myers 1952; Rossel et al. 2002; Schuster et al. 2004; Schuster et al. 2006; Temple 2007; Timmermans 1975; Timmermans 2000; Timmermans 2001; Timmermans & Maris 2000; Timmermans & Souren 2004; Timmermans & Vossen 2000). The diet of *T. chatareus* was studied in two tropical mangrove estuaries in northern Australia (Blaber 2000) and it was reported that their diet includes crabs, shrimps, insects, and plant materials.

However, despite the feeding techniques, and concise information about the diet, no quantitative data could be found on the diet of *T. chatareus*, *T. jaculatrix* or any toxotid in Malaysia or elsewhere. Besides, no information on their gastric emptying time or the rate at which food passes from the stomach, digestion and nutrient absorption along the alimentary tract are available in the literature.

The taxonomy of archer fishes is very perplexing because of their similar morphological features; consequently, these two species (*T. chatareus* and *T. jaculatrix*) are identified erroneously and named by the local people 'Sumpit'. At present, detailed biology of these fishes is scarce. One of the major obstacles to thorough biological data still remains the difficulty associated with the correct taxonomic identification of tropical marine and estuarine fishes (such as archer fishes) (Blaber 2000). Furthermore, archer fishes *T. chatareus* and *T. jaculatrix* are relatively scarce and specimen collection is cumbersome within a complex rooting system of mangrove forest. Their sharp eye vision and fast swimming speed make it further difficult. The lack of information about the life history of archer fishes is partly caused by the complexity of sample collection.

The small size and attractive colouring of archer fishes (especially *T. chatareus* and *T. jaculatrix*) have made them very popular in the aquarium trade therefore, another noticeable threatening aspect of these fishes is 'life specimens' collection for aquarium trade. The coastal waters of Johor, located in the southern region of peninsular Malaysia, have experienced extensive habitat destruction crucial for archer fishes, primarily mangrove vegetation. This destruction is attributed to the utilization of destructive and indiscriminate fishing techniques, including the illegal use of fishing gears, occasional poisoning, coastal zone reclamation, effluent discharge, and overall environmental degradation by both commercial and artisanal fishers. Archer fishes are able to withstand man-made ecological pressure and natural hazards, successfully surviving in such menacing environments. The archer fishes are good example of fish species with high degree of resilience. Resiliency is the ability of animals to undergo, absorb and respond to change and disturbance while changing its function.

They inhabit the interface between land and sea at low latitude mangroves, which can be described as a harsh environment, daily being subjected to tidal changes in temperature, water and salt exposure, and varying degrees of anoxia. Mangrove forests and their associated animals are therefore fairly robust and highly adaptable (or tolerant) to live in waterlogged saline soils within tropical seascapes. Therefore, it exhibits a high degree of ecological stability or resilience.

Data on fishes with high degree of resilience is still lacking. There are many fields of research related to resilience of marine organisms with most likely similar objectives and outcome. The high degree of resilience in fish species such as archer fish can be determined through feeding habit and trophodynamic analysis, morphological characteristics, ageing and reproductive potential of these fishes. Resilience of these fishes and other resilient species is awaiting further critical analysis. Indeed, a further description of the resiliency of archer fish can be used as baseline information for other similar resilient species in the mangrove ecosystem.

The importance of this book lies on its specific focus on studies of a number of important biological features e.g., morphometry, age, growth, condition factors, feeding habits, trophic level, gastric emptying time, digestion, nutrient absorption along the alimentary tract and reproductive potential of two archer fishes. Thus far this book will be the pioneer discussion on the biology of *T. chatareus* and *T. jaculatrix* or any toxotid in Malaysia or elsewhere. The chapters discussed will expand the literature for these fishes and the information can be used as a baseline data for future comparison.

Biology of Archer Fish

Archer fish are included in the Order Perciformes, Family Toxotidae. The order Perciformes, the largest group of vertebrates, encompasses 22 suborders of fish, about 1,367 genera and over 7800 species (Nelson 1984). This order is defined by their laterally compressed bodies and ctenoid scales. Typically, they have one or two dorsal fins but never an adipose fin. The dorsal and anal fins are divided into anterior spiny and posterior soft-rayed portions, which may be partially or completely separated. The pelvic fins usually consist of one spine and up to five soft rays, positioned unusually far forward near the throat or underneath the belly (Figure 1.1). There are four to seven branchiostegal rays as well as gill rakers and toothed. Moreover, the order Perciformes has 24 or more vertebrae (Craig 1987).

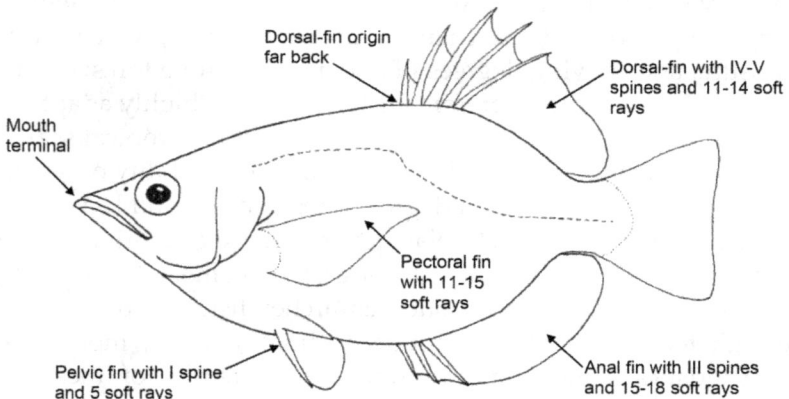

Figure 1.1. Common characteristics of family Toxotidae.
Source: Allen 2001

In comparison with other species in the genus *Toxotes*, *Toxotes chatareus* has larger scales; 5 dorsal spines; 5 to 7 black blotches on upper sides; 30 to 36 lateral line scales. Therefore, this species is also known as large scale and spotted archer fish. The body of *T. chatareus* is oblong and compressed; head flattened on dorsal surface; 5 dorsal spines and 30 to 36 lateral line scales. Body colour generally pale (grey to silvery); a series of 5 to 7 black blotches on upper sides; dorsal and anal fins dusky to blackish; pectorals, pelvic and caudal fin slightly dusky (Figure 1.2). Maximum total length of this species about is 50 cm and very rarely it exceeds 35 cm (Allen 2001).

Figure 1.2. External view of *Toxotes chatareus*.

Toxotes jaculatrix has 4 dorsal fin spines, 26 to 30 lateral line scales, and a series of 4 or 5 black bars on upper sides. This species is also known as banded archer fish. The body is oblong and compressed; head is flattened on dorsal surface. Dorsal and anal fins of this species are dusky to blackish; pectorals, pelvic and caudal fin usually pale (Figure 1.3). Maximum total length is about 30 cm, commonly to about 20 cm.

Species Identification

Comparing anatomical features of organisms has been the basis for taxonomic classification of organisms and differentiation among stocks as well as closely related species. Those comparisons were traditionally carried out using differences in body measurements (morphometry) or differences in numbers of anatomical structures (meristics) (Bagenal 1978; Bookstein 1991; Dryden and Mardia 1998).

Figure 1.3. External view of *Toxotes jaculatrix*.

From the sixties till until now, morphometrics was focused on the application of multivariate statistical tools to morphological variables; in order to describe shape variation within and between groups as well as in species and, this approach is referred as 'multivariate morphometrics' (Blackith and Reyment 1971; Bolles and Begg 2000; Dryden and Mardia 1998; Tudela 1999). Multivariate analysis of morphometric and meristic characters is a standard tool for defining population units and differentiating between genera, species, sub-species, and groups of animals (e.g., Boetius 1980; Bolles and Begg 2000; Fridriksson 1958; Pierce et al. 1994; Thorpe 1976; Tudela 1999). Meristic characters have enumerable morphological features such as fin rays, gill rakers and vertebrae, whereas morphometric characters are those obtained by measurements of body parts. Morphometric and meristic differences can arise when genetic isolation allows genotypic differences to arise through local differences in selective pressures, mutation or genetic drift (Hadon and Willis 1995; Mamuris et al. 1998).

T. chatareus and *T. jaculatrix* are very similar in appearance and are somewhat difficult to distinguish from each other. Allen (2001) and Day (1971) described some differences in the number of soft rays in the anal fin between *T. chatareus* and *T. jaculatrix*. Number of dorsal spines and body colouration are currently used to differentiate *T. chatareus* and *T. jaculatrix* (Allen 2004). However, colour is one of the most variable characteristics, capable of changing in different geographic areas, depths, and in response to variations in food composition. Multivariate analyses of morphometric and meristic characters might play a role to differentiate these closely related species more precisely along with Allen's description.

Age and Growth

Information on age, asymptotic length, growth coefficient, and growth performance is essential for efficient sustainable management of fish resources. Numerous methods have been used to determine the age of fishes including counting the number of annually formed growth zones in hard tissues, such as scales, otoliths and less frequently certain other cartilaginous or bony structures (Adeyemi 2009; Bagenal and Tesch 1978; Bilecenoglu 2009; Campana 2001; Campana and Neilson 1985; Zorica et al. 2005). These growth zones are formed in response to seasonal variations in environmental conditions, leading to changes in the deposition of protein and calcium during alternating slow and fast phases of growth (Campana and Neilson 1985).

In temperate fishes, thin growth zones typically develop during the cooler 'winter' months, while thick zones form over the warmer 'summer' months in their otoliths (Booth and Buxton 1997; Francis et al. 1992; Johnson 1983; Sarre and Potter 2000). On the other hand, aging studies of fishes from tropical waters have reported annual increments in calcified structures like scales (Chung and Woo 1998; Mayekiso and Hecht 1988; Werder and Soares 1985), dorsal and pectoral spines (Ezenwa and Ikusemiju 1981; Pantulu 1961), vertebral centra (Brown and Gruber 1988), and otoliths (Fowler and Doherty 1992).

As numerous studies have shown that the use of scales for ageing teleosts often yields unreliable results (Beamish and McFarlane 1987; Casselman 1987), otoliths are now almost exclusively used for ageing bony fishes. However, when using whole otoliths to age fish, the ages of older fish can be underestimated because of the difficulties in detecting the outer growth rings as a result of the allometric growth of the otolith (Beamish and McFarlane 1987; Casselman 1987; Hyndes et al. 1992). Thus it is often necessary to section otoliths in order to reveal all of their growth zones. As the sectioning of otoliths is time consuming, workers often make preliminary comparisons between the results obtained by counting growth zones on both whole and sectioned otoliths of a wide size range of fish to ascertain whether it is necessary to section all of the otoliths or just those with the most annuli (Hesp et al. 2002; Hyndes et al. 1992; Hyndes et al. 1998; Sarre and Potter 2000). If there are discrepancies between the numbers of zones in whole and sectioned otoliths, conducting such comparisons allows us to determine the specific point at which these discrepancies begin to occur. Then ages of small and young fish can often be determined accurately by using the number of annuli on their whole otoliths, whereas ages of larger and older fish have to be determined using sectioned otoliths (Campana 1984).

A major requirement of all ageing studies is that the growth zones used for ageing are validated as having been formed annually (Beamish and McFarlane 1983; Campana 2001; Casselman 1987; Francis et al. 1992; Hyndes et al. 1992). In the past, the failure to carry out age validation on many occasions has led to inaccurate ageing of fish and thus to such a gross misunderstanding of the biology of some species that fisheries managers did not establish appropriate plans to prevent serious overexploitation of fish populations or species (Campana 2001). One method of validating that growth zones are formed annually in hard structures is by labeling the calcified tissue, which involves the capture, tagging and then injection of a label, e.g., tetracycline, and the subsequent recapture of those fish. Since the labeling compound leaves a mark in the calcareous structure, the use of the number of zones as annuli in those structures for ageing a particular species is valid if the number of zones in the otolith peripheral to the mark corresponds to the number of years between releases and recapture (Casselman 1983; Casselman 1987).

Another technique for validating that the growth zones on the calcified tissue of fish are formed annually is marginal increment analysis (Campana 2001; Hyndes et al. 1992). The marginal increment is the distance outside the outermost (opaque) zone. If the opaque and translucent zones in otoliths are formed on an annual basis, there should be a significant decrease in the marginal increment at a specific time of the year. This occurs when a newly formed opaque zone starts to recede and a fresh translucent zone begins to form at the edge of the otolith. The marginal increment should then gradually increase as the translucent zone widens during its continued formation (Hyndes et al. 1992). Due to its modest sampling requirements and relatively low cost, this is the most commonly used method for age validation. However, Campana (2001) stated that marginal increment analysis is also one of the most difficult ageing validation methods to be carried out properly. This is because of the difficulties associated with viewing a partial increment affected by variable light refraction, through an edge which becomes increasingly thin as the margin is approached, as well as by light reflection off the curved surface of the edge (Campana 2001). However, this approach is valid if carried out with sufficient rigour (Campana 2001).

Population growth is another important demographic parameter. The study of growth means basically the determination of the body size as a function of age. Therefore, all stock assessment methods work essentially with age composition data. Pütter (1920) developed a growth model which can be considered the base of most other models on growth including the one developed as a mathematical model for individual growth by von Bertalanffy (1938), and which has been shown to conform to the observed growth of most fish species. The theory behind various growth models has been reviewed by notable researchers such as Beverton and Holt (1957), Ursin (1968), Ricker (1975), Gulland (1983), Pauly (1984), and Pauly and Morgan (1987). However, it is important to note that the present study focuses solely on the von Bertalanffy growth model, which relates body length to age. The equation used in this study is represented as $L_t = L_\infty(1 - \exp{-K^{(t-t0)}})$, where L_t represents the total length (cm) length at age t (years), L_∞ denotes the predicted asymptotic length (cm) derived from the equation, K represents the growth coefficient (year^{-1}) and t_0 is the hypothetical age (years) at which fish would have a length of zero if growth followed the predicted pattern (Punt and Hughes 1992; Ricker 1975; von Bertalanffy 1938).

Additionally, population growth can also be estimated through length-weight relationship (LWR) analysis. Length-weight relationships are the two basic components in the biology of fish species at individual and population levels (Kohler et al. 1995). These relationships are useful in fishery management for both applied and basic use (Pitcher and Hart 1982) to: (i) estimate weight from length observations; (ii) calculate production and biomass of a fish population; and/or (iii) provide information on stocks or organism condition at the corporal level. Although LWR is readily available for most European and North American freshwater and marine fishes (e.g., Bagenal and Tesch 1978; Koutrakis and Tsikliras 2003; Leunda et al. 2006; Miranda et al. 2006; Oscoz et al. 2005; Petrakis and Stergiou 1995; Sinovcic et al. 2004), adequate local information is still scarce for most tropical and subtropical fish species (Ecoutin et al. 2005; Harrison 2001; Martin-Smith 1996). Condition factor (Fulton condition factor K, and relative condition factor K_n) is a quantitative parameter of the state of well-being of the fish that determines present and future population success by its influence on growth, reproduction and survival. The condition of a fish reflects recent physical and biological circumstances, and fluctuates with interaction among feeding conditions, parasitic infections and physiological factors (Bolger and Connolly 1989; Le Cren 1951; Ney

1993). In less abstract terms, it means that between two fishes of the same length, the heavier one will be in a better condition (Bagenal and Tesch 1978; Ricker 1975). The seasonal changes in this index, which typically exhibit minimum values shortly after spawning, have been frequently observed (Jobling 1993).

At present no information is available on age, population growth and condition factors of *T. chatareus*, *T. jaculatrix* or any toxotid, in Malaysia or elsewhere. Consequently, lack of adequate knowledge about the growth, conditions and ageing of *T. chatareus* and *T. jaculatrix* remains an impediment to the formulation of sound management strategies for these two interesting fishes.

Feeding Habits

Archer fishes exhibit nature's most amazing feeding technique i.e., they are capable of bringing down aerial insect prey from the overhanging vegetation with a precisely aimed jet of water (Lüling 1958, 1963). The primary focus of attention regarding this unique method of prey catching revolved around two key inquiries: how the fish ejects a squirt of water (Elshoud and Koomen 1985), and how it can hit its target despite the shift of the prey's image caused by the refraction of light on the surface of the water (Dill 1977).

Lüling (1958, 1963) assumed that the greater part, the fish would avoid both problems by squirting nearly vertically. However, this idea is contradicted by the Verwey (1928) report, Lüling's own illustrations (Lüling, 1958, 1963, 1964, 1969), and by the Herald (1965) photograph, all showing *T. jaculatrix* squirting at angles between 70° and 80°. Bekoff and Dorr (1976) did not provide support for Lüling's theory, as their findings revealed that targets positioned at a height of 30 cm were frequently missed by a margin of 15-20 cm. Moreover, Timmermans and Vossen (2000) reported *T. chatareus*, and probably *T. jaculatrix* often use aiming angles other than 90°. The aiming angle is the angle between the water surface and the line from the fish's snout to the target, at the moment of squirting. Lüling (1958) reported that most misses scored by *T. jaculatrix* were on the azimuth (left or right of the target) not in elevation (over or under the target).

Archer fish also able to catch aerial prey by means of jumping, but the success rate depends on the prey's height from the water level. Timmermans (2001) reported that archer fish are successful about one body length while Verwey (1928) reported that they are successful about maximally 14 cm. On the other hand, Lüling (1958), documented that they are successful within few cm but do squirt as high as 20 times their body length (Timmermans 2000; Verwey 1928). However, they also affirmed that jumping cannot serve to evade the refraction effect. To date detailed studies on spitting success rate and accuracy in different height ranges of archer fishes in group and solitarily are scarce. The majority of studies have concentrated on aiming angle, refraction, and spitting behaviour of archer fish (Milburn and Alexander 1976; Myers 1952; Rossel et al. 2002; Schuster et al. 2004; Timmermans and Souren 2004; Wöhl and Schuster 2006).

For scientific study, fisheries management, or conservation reasons, an understanding of the diets, feeding ecology and trophic interrelationships of estuarine fishes is fundamental. However, very few published reports are available on the feeding studies of archer fishes. Blaber (2000) studied the diet of *T. chatareus* at two tropical mangrove estuaries in northern Australia. He noticed ontogenetic changes in the diet of *T. chatareus* e.g., small fish (<100 mm) eat no plant material, medium-sized fish (100-139 mm) ingested some plant material, while about a quarter of the intake of larger fish consists of parts of mangrove trees. He pointed out that in medium and large fish, these are to some extent replaced by larger insects such as Orthoptera. However, overall, with increasing fish size, arboreal insects are replaced by larger arboreal prey, particularly grapsid crabs, which make up about half of the diet across all size groups.

A small proportion of the diet of all size groups consists of animals captured in the water, such as amphipods, carids, penaeids and fish (Blaber 2000). The increasing incidence of plant material in the diet is unusual; it is not digested and its intake could be accidental. The absence of mangrove material in small fish may be because their jet of water is insufficient to break off the plant material. Either of these scenarios suggest that the seizing of prey by the archer fish is a reflex action to material falling onto the water, and that it may be unable to discriminate between living and inanimate material (Blaber 2000).

It is possible that the plant material is consumed deliberately, perhaps in order to supplement the diet in some unknown way or to aid in the digestion of arthropods. It is possible that the shooting method may also be employed for aquatic prey as suggested by Elshoud and Koomen (1985). Regardless of the answers to these questions, it is apparent that this mode of feeding can only be employed successfully in quiet water areas with overhanging vegetation with epifauna-conditions such as those existing in the mangrove forests of the middle and upper reaches and tributary creeks of tropical estuaries.

Despite all of the above studies, no quantitative data could be found on the diet of any of the seven species of *Toxotes*. But indeed, the feeding preferences of fish species are important in classic ecological theory, mainly in identifying feeding competition (Bacheler et al. 2004), structure and stability of food webs (Post et al. 2000), omnivory (Pimm and Lawton 1978) and assessing predator-prey functional responses (Dorner and Wagner 2003). Additionally, the key role of feeding studies for fisheries biology and ecology and, more importantly, for fisheries management, was uncovered only in the last decade with the use of trophic level (TROPH) in predicting the effects of fishing on the balance of marine food webs (Pauly et al. 1998). The trophic level, which for marine animals' ranges between 2 (for herbivores/ detritivores) and 5.5 (for specialised predators of marine mammals), expresses the relative position of an animal in the food web that nourishes them (Pauly et al. 2000).

The extraordinary array of feeding adaptations and mechanisms in archer fishes allows them to exploit almost all available sources of food (Norman & Greenwood 1975). Though morphology can provide circumstantial evidence of the diet of a fish, the inferences must be confirmed by more direct evidence of what is eaten (Wootton 1990). A description of the composition of the diet should indicate the relative importance of the items eaten. Diet can rarely be studied by directly observing feeding behaviour and identifying what is eaten (Wootton 1990). Usually diet is sampled by extracting the stomach contents, either by killing the fish and dissecting the stomach or by flushing out the stomach. Several methods are used to provide a quantitative description of such samples (Hynes 1950; Hyslop 1980; Langler 1956; Pillay 1952; Windell 1971); probably no one method is entirely satisfactory. The simplest method estimates the frequency of occurrence in stomachs (Hyslop 1980). After identification of the food categories present, the number of stomachs in which a given category is found is expressed as percentage of the total number of stomachs sampled. To comprehend the importance of the topic, Blaber (2000) stated that much further research is needed on trophic level for tropical estuarine fishes (such as archer fishes), but a sound knowledge of their diets is an essential prerequisite for a deeper understanding of their feeding ecology and behavior.

Gastric Emptying and Digestion

The terms digestion rate and gastric evacuation rate are often used interchangeably to denote the rate at which food passes from the stomach into the intestine even though this is something of a misnomer (Windell 1978). Knowledge of fish gastric evacuation rates (GERs) is a necessary component for both field and laboratory studies concerned with fish feeding rates, energy budgets, and the trophic dynamics of aquatic systems (Forrester et al. 1994; Sweka et al. 2004). In fisheries, research using consumption models has depended more heavily on rates of gastric evacuation than research in other fields, such as animal nutrition, which often focus more on the rate of transit through the entire gastrointestinal track (Butler et al. 2004; Dorcas et al. 1997; Henriques et al. 2004; Roxburgh and Pinshow 2002; Sponheimer et al. 2003).

In general, the alimentary tracts of fish are much simpler than the alimentary tracts of other animals, such as mammals (Stevens & Hume 1995) and food consumption estimates are usually determined through direct gut content analysis using lavage techniques or by sacrificing large numbers of animals (Adams & Breck 1990). Conversely, estimating food consumption in many animals requires indirect methods, such as inferring feeding rates of wild animals from feeding rates of captive animals (Innes et al. 1987) or through bioenergetic modeling, because lavaging certain types of animals or sacrificing large numbers of those animals may be logistically difficult, prohibited, or simply undesirable (Winship et al. 2002).

Gastric emptying (or evacuation, GE) has been measured in several ways. It is not always easy to compare the results from different authors who used different methods because of the differences in specific objectives and fish used in the experiments. Even different methods of handling have shown different results. In the laboratory, GE is often estimated by giving a group of experimental fish a meal of known weight or volume and measuring the amount remaining in the stomach of individual fish by serial slaughter at successive intervals after feeding (also known as force feeding/gastrectomy) (Basimi and Grove 1985a; Fletcher 1984; Fletcher et al. 1984; Jones 1974; Mazlan 2001; Swenson and Smith 1973; Windell 1966). This method has certain advantages since it provides direct observation of food items in the stomach. However, large numbers of fish are required in the experiment, and the described GE process only represents the population average, largely ignoring individual variation. This method allows estimation of Gastric Emptying Time (GET, usually in hours) and also Gastric Emptying Rate (GER, usually in grams per hour); if the gastric emptying curve is not linear, GER varies as digestion proceeds. Direct observations in the field have been attempted in a similar way by observing the stomach contents of sampled fish at different times after capture (Andersen and Beyer 2008; Basimi and Grove 1985b; Elliott 1973; Elliott and Persson 1978; Seyhan and Grove 1998; Thorpe 1977).

When the numbers of experimental fish are limited, many workers prefer to use an X-ray technique to facilitate direct observation without killing the fish. This method has been applied regularly in laboratory studies in the last three decades and is widely practiced but with some modifications in equipment and processing between authors (Hossain et al. 1998; Hossain et al. 2000; Mazlan and Grove 2003; Molnár and Tölg 1960, 1962; Talbot 1985). Originally, the process of digestion was monitored by using X-ray-dense structures in the food (otoliths, bones, chitin) (Molnár and Tölg 1960, 1962) but inclusion of exogenous marker (e.g., barium sulphate, $BaSO_4$) allowed digestion of soft-bodied prey such as polychaetes to be monitored (Edwards 1971). Barium sulphate as a dispersed contrast medium has been widely used (Edwards 1973; Elliott 1972; Jobling et al. 1977; Mazlan and Grove 2003; Ross and Jauncey 1981; Seaburg and Moyle 1964). This was done by mixing radio-opaque $BaSO_4$ at a concentration of around 20-25% w/w with the experimental food, to produce high resolution X-ray images (Edwards 1971; Edwards 1973; Grove et al. 1978; Grove et al. 1985; Jobling et al. 1977; Mazlan 2001; Mazlan and Grove 2003).

Meal composition, or energy density, can vary greatly with prey type and prey size, thereby having another very important influence on consumption and digestion rates. Many studies on fish have found that meals high in energy result in an increase in time to 100% gastric evacuation (Flowerdew and Grove 1979; Hopkins and Larson 1990). Additionally, diets with an added diluent, a non-digestible marker that lowers a meal's energy density, were evacuated more rapidly from fish stomachs than those with higher energy content, thereby exhibiting an increased digestion rate (Flowerdew and Grove 1979; Jobling 1980a). High-energy fats tend to slow gastric emptying more than proteins or carbohydrates (Jobling 1980a). Therefore, mature prey fish which are high in lipid content will slow digestion compared to juvenile prey fish, which may be high in protein, but low in lipids. While generally lipid-rich mature preys are physically larger than immature preys of the same species, any increase in prey size decreases the surface area to volume ratio available for enzymatic digestion (Swenson and Smith 1973). Therefore, not only do mature individuals contain more lipids, which slow gastric evacuation rates, but they also have a lower surface area to volume ratio which slows gastric evacuation rates even further.

Linear, exponential, and square root gastric evacuation models have commonly been used to quantify the gastric evacuation processes of fish species, and then are either input directly into, or used to meet the assumptions of a consumption model (Andersen 1998, 1999, 2001; Andersen and Beyer 2005; Berens and Murie 2008; Brodeur 1984; dos Santos and Jobling 1992; Hall et al. 1995; Hopkins and Larson 1990; Jobling 1980a, 1981b; Jobling and Davies 1979; Temming and Andersen 1994; Salvanes et al. 1995). A linear model often describes the gastric evacuation processes of top carnivores, or piscivores, which tend to consume only a few fairly large prey items over a feeding cycle, thereby causing relatively long digestion times compared to the length of their feeding period (Adams and Breck 1990). Jobling (1987) attributes the linear model to large food particles with lower surface-to-volume ratios, low fragmentation rates, and high dietary energy densities. These types of prey items tend to be evacuated from the stomach at a constant rate. Previous work with piscivores and linear digestion processes have included studies on black and yellow rockfish *Sebastes chrysomelas* (Hopkins and Larson 1990), plaice *Pleuronectes platessa* (Jobling 1980b), and walleye *Stizostedion vitreum vitreum* (Swenson and Smith 1973).

In general, the alimentary tract or canal of fish is divided into different functional parts, namely the mouth and buccal cavity, the pharynx, the oesophagus, the stomach, the intestine and its associated organs, and the rectum (Figure 1.4). Most post-larval fish develop a stomach but stomachless fish have evolved in separate families. The alimentary canal structure within a family is strongly associated with feeding habits, mode of feeding and the ecological niche of the species (Fänge and Grove 1979; Jobling 1995a; Kapoor et al. 1975; Smith 1989).

Digestion is the act of mechanical and enzymatic breakdown in the fish's stomach that converts food into soluble and diffusible products capable of being absorbed, or assimilated, by cells in the fish's stomach and intestine (Knutsen and Salvanes 1999). The digestive and absorptive properties of the alimentary canal of fishes are very important in determining rate of digestion, daily rations, gastric emptying/evacuation etc. (Weatherley and Gill 1987). The characteristics of the alimentary canal determine the extent to which dietary nutrients are digested and assimilated.

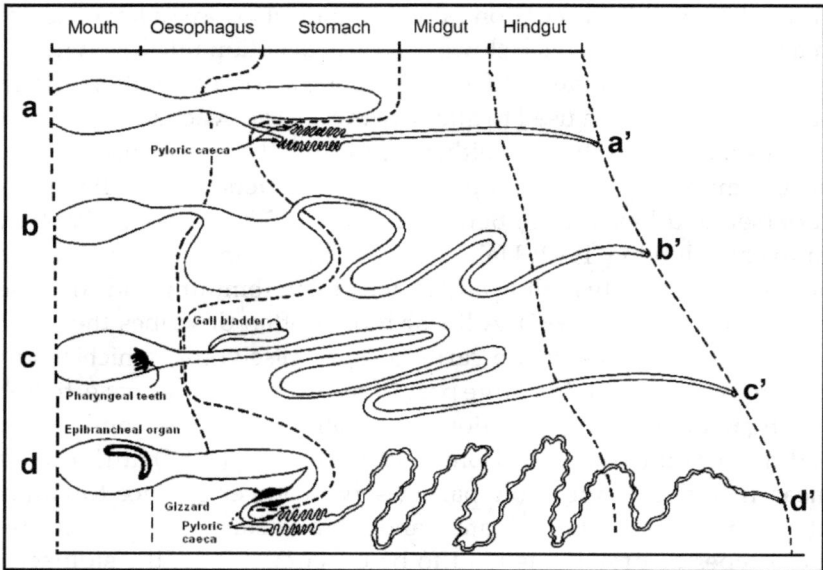

Figure 1.4. General morphology of the alimentary tract/canal in fish. (Intestine = combination of mid gut and hindgut). (a-a', b-b', c-c', d-d' indicate entire alimentary tract of different fishes). The archer fish has a structure like a-a' but the intestine after the pyloric caeca more closely resembles b-b'. *Source:* Smith 1989.

Food materials are usually broken down in the stomach of carnivorous fish through a combination of muscular contractions and enzymatic action in the acidic medium. The breakdown products are expelled from the stomach through the pyloric sphincter as chime into the anterior intestine through the process known as gastric evacuation. Many species have prominent appendages on the anterior intestine; the finger-like pyloric caeca (intestinal caeca), which are of various types depending on the fish species. These blind-ended tubes serve to increase intestinal absorptive surface area without increasing the length or thickness of the digestive tract (Buddington et al. 1987; Buddington and Diamond 1987). They can be considered as an extension of the small intestine with similar digestive and absorptive functions (Bergot et al. 1981; Boge et al. 1979; Infante and Cahu 1994).

Dimes and Haard (1994) used pyloric caeca of salmonid (*Salmo gairdneri*) to monitor protein digestion and absorption ('digestibility') and noted that pH-static methods provide good correlates of protein digestibility found by conventional methods (analysis of food and of the faeces). Rønnestad et al. (2000) found that retrograde peristaltic contractile activity in the pyloric region provides a mechanism by which the fish can fill the caeca with chime. Observations on archer fish digestive tracts show densely-packed pyloric caeca which are likely to be an important site for nutrient absorption

Reproductive Biology

Reproductive development and reproductive histology in male and female are well understood by histological technique. Histology is the most detailed and accurate method for assessing gonad maturity stages of fishes (West 1990). Developmental stage of the gonads (or germ cells) is an endpoint commonly required from histological analysis in monitoring programs and other reproductive studies. Consequently, plenty of research has been conducted and scientists have affirmed different developmental stages of fish gonad. Yamamoto and Yamazaki (1961) describe ten stages, Treasurer and Holliday (1981) nine stages, Braekevelt and Mc Millard (1967) eight stages, Borg and Van Veen (1982) seven stages, Htun-Han (1978), Solomon and Ramnarine (2007) and Arocha and Bárrios (2009) six stages, Rideout et al. (1999) and Srijunngam and Wattanasirmkit (2001) five stages and Goodbred et al. (1997), Yön et al. (2008) and Mazlan and Rohaya (2008) four stages.

As with most aspects of fish ecology, much of the information about reproduction of fishes comes from studies of temperate species (Blaber 2000). The reproductive biology of the vast majority of tropical species has never been studied, and inferences drawn from temperate species may, at best, be misleading (Blaber 2000). However, reliable data on the reproductive biology of fish species is essential for effectively managing the fisheries for those species. For example, knowledge of the spawning period of a species can facilitate the protection of its stocks by imposing bans on their capture when they are spawning or through the permanent or semi-permanent closure to fishing of areas in which spawning aggregations occur (Jennings et al. 2001; Jones et al. 2002; Sadovy 1996). Such knowledge is also required to enable the acquisition of data from that time of year for determining the length at maturity of a species, which is commonly used in setting an appropriate minimum legal length for capture (Beverton and Holt 1957; Hill 1990; Sadovy 1996; Winstanley 1990). Furthermore, elucidation of whether a species is a single or multiple spawner and the estimation of its fecundity is dependent on data derived from samples collected during its spawning period (Hunter et al. 1992).

Nevertheless, relatively little is known about the reproductive biology of archer fishes. Allen (1991, 2002) stated that *T. chatareus* and *T. jaculatrix* breed both in fresh and brackish waters and their breeding takes place in the wet season. Pethiyagoda (1991) described the fecundity (20,000 to 150,000) and egg sizes (0.4 mm) of *T. chatareus*. Other than that, no data can be found in the literature. Consequently, lack of adequate knowledge on the reproductive biology of *T. chatareus* and *T. jaculatrix* remains an impediment to the formulation of sound management strategies for these fascinating fishes.

Morphometric and Meristic Variation of Archer Fish

Interspecies variation for fishes and other aquatic animals based on morphometric analyses have been reported elsewhere. However, no information on morphometric and meristic analysis of *Toxotes chatareus*, *Toxotes jaculatrix* or any toxotid has been documented in Malaysia or elsewhere. The genus *Toxotes*, in particular of the species *Toxotes chatareus* and *Toxotes jaculatrix* are very similar in appearance and until now external marking (body colouration) and/or number of dorsal spines (simple descriptive statistics) are being used to differentiate these two species (Allen 2001, 2004).

The aim of this chapter is to find out other morphological distinctions that can also be used to differentiate these closely related species more precisely along with Allen's description.

The Approach

A total of 128 fish (*T. chatareus*, n = 63; *T. jaculatrix*, n = 65) were frozen soon after collection and defrosted for laboratory analyses, which took place about 2 months later to ensure that all fish were analysed following a similar period of freezing. Digital photographs were taken on the left side of each fish with a Canon Power Shot A640 (10.0 mega pixels; Canon, Tokyo, Japan) camera, and body colouration was recorded for species identification (Figure 2.1a, b).

In the laboratory, a total of 31 morphometric measurements were recorded for each fish. Measurements generally followed Allen (2004) and Pouyaud et al. (1999). Descriptions of the 31 morphometric characters are given in Table 2.1 and illustrated in Figure 2.2. Morphometric measurements were taken from the left lateral aspect and measured to the nearest 0.01 cm using a digital caliper (absolute digimatic digital calipers, Mitutoyo, Singapore).

The number of vertebrae (TV) were counted from Microradiographic unit (Figure 2.1c) taken with a microradiographic unit (M60, Softex, Tokyo, Japan). A stereo microscope (Stemi DV4/DR, ZEISS,

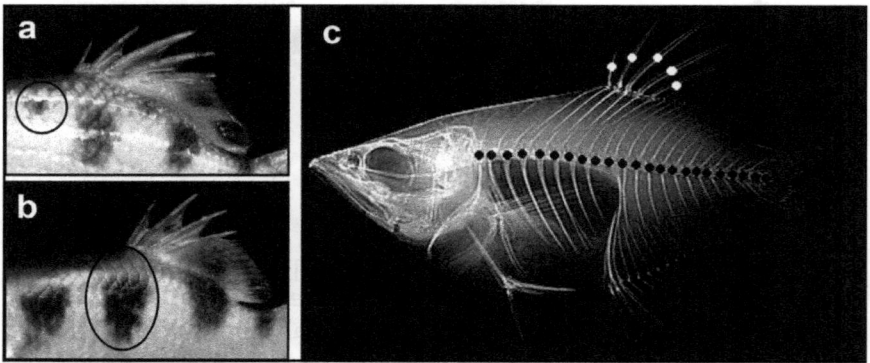

Figure 2.1. Differences in body colouration between *Toxotes chatareus* (a) with 6-7 alternating vertical black bars and black spots and *Toxotes jaculatrix* (b) with 4-5 vertical black bars. Radiographic image of *T. chatareus* (c) showing number of vertebrate (black spots) and number of dorsal fin spines (white spots)

Oberkochen, Germany) was used for counting fin rays. All meristic characters were counted twice on the same day by the same observer. The number of dorsal fin spines (DS) were excluded from the present study because this trait is constant in the two species (*T. chatareus*, DS = V; and *T. jaculatrix*, DS = IV) as documented by Allen (2004).

Box plot analysis was performed to examine the distribution and to detect the presence of extreme outliers. Extreme outliers are defined as those which lie more than three times the inter-quartile range to the left or right from the first and third quartiles respectively. Separate statistical analyses were conducted on morphometric and meristic data since morphometric data are continuous and more susceptible to the environmentally induced variability while meristic data are discrete and fixed early in development (Hermida et al. 2005; Ihssen et

Table 2.1. Definitions of morphometric measurements and meristic counts of *Toxotes chatareus* and *Toxotes jaculatrix* used in this study

Characters		Acronyms
31 Morphometric measurements		
Standard length	: Tip of the upper jaw to the tail base	SL
Pelvic fin length	: From base to tip of the pelvic fin	PVFL
Pelvic spine length	: From base to tip of the pelvic spine	PVSL
Pectoral fin length	: From base to tip of the pectoral fin	PFL
Caudal fin length	: From tail base to tip of the caudal fin	CFL
Pre-dorsal length	: Front of the upper lip to the origin of the dorsal fin	PDL
Pre-anal length	: Front of the upper lip to the origin of the anal fin	PAL
Pre-pectoral length	: Front of the upper lip to the origin of the pectoral fin	PPCL
Pre-pelvic length	: Front of the upper lip to the origin of the pelvic fin	PPVL
Length of dorsal fin base	: From base of first dorsal spine to base of last dorsal ray	LDFB
Length of anal fin base	: From base of first anal spine to base of last anal ray	LAFB
First dorsal spine length	: From base to tip of the first dorsal spine	LDS1
Second dorsal spine length	: From base to tip of the second dorsal spine	LDS2
Third dorsal spine length	: From base to tip of the third dorsal spine	LDS3
Fourth dorsal spine length	: From base to tip of the fourth dorsal spine	LDS4
Fifth dorsal spine length	: From base to tip of the fifth dorsal spine	LDS5
First anal spine length	: From base to tip of the first anal spine	LAS1
Second anal spine length	: From base to tip of the second anal spine	LAS2
Third anal spine length	: From base to tip of the third anal spine	LAS3

(Contd.)

Table 2.1. (*Contd.*)

Characters		Acronyms
Length of soft dorsal ray	: From base to tip of the soft dorsal ray	LSDR
Length of soft anal ray	: From base to tip of the soft anal ray	LSAR
Mouth height	: Maximum vertical measurement of the mouth when completely open	MH
Upper jaw length	: Straight line measurement between the snout tip and posterior edge of maxilla	UJL
Lower jaw length	: Straight line measurement between the snout tip and posterior edge of mandible	LJL
Length of caudal peduncle	: From base of the last anal fin ray to middle of caudal fin fold	LCP
Body depth	: Maximum depth measured from the base of the dorsal spine	BD
Body width	: The greatest width just posterior to the gill opening	BW
Snout length	: The front of the upper lip to the fleshy anterior edge of the orbit	SNL
Eye diameter	: The greatest bony diameter of the orbit	ED
Head length	: From the front of the upper lip to the posterior end of the opercular membrane	HL
Depth of caudal peduncle	: The least depth of the tail base	DCP
9 Meristic counts		
Dorsal fin ray	: Number of soft fin rays in dorsal fin	DFR
Anal fin ray	: Number of soft fin rays in anal fin	AFR

al. 1981; Turan et al. 2006). Spearman's rank correlation indicated that there was low association between meristic characters and standard length (SL) of the samples so the meristic characters were not adjusted for size differences. In contrast significant Pearson's correlations were observed between size and morphometric characters and allometric effects may accentuate such differences.

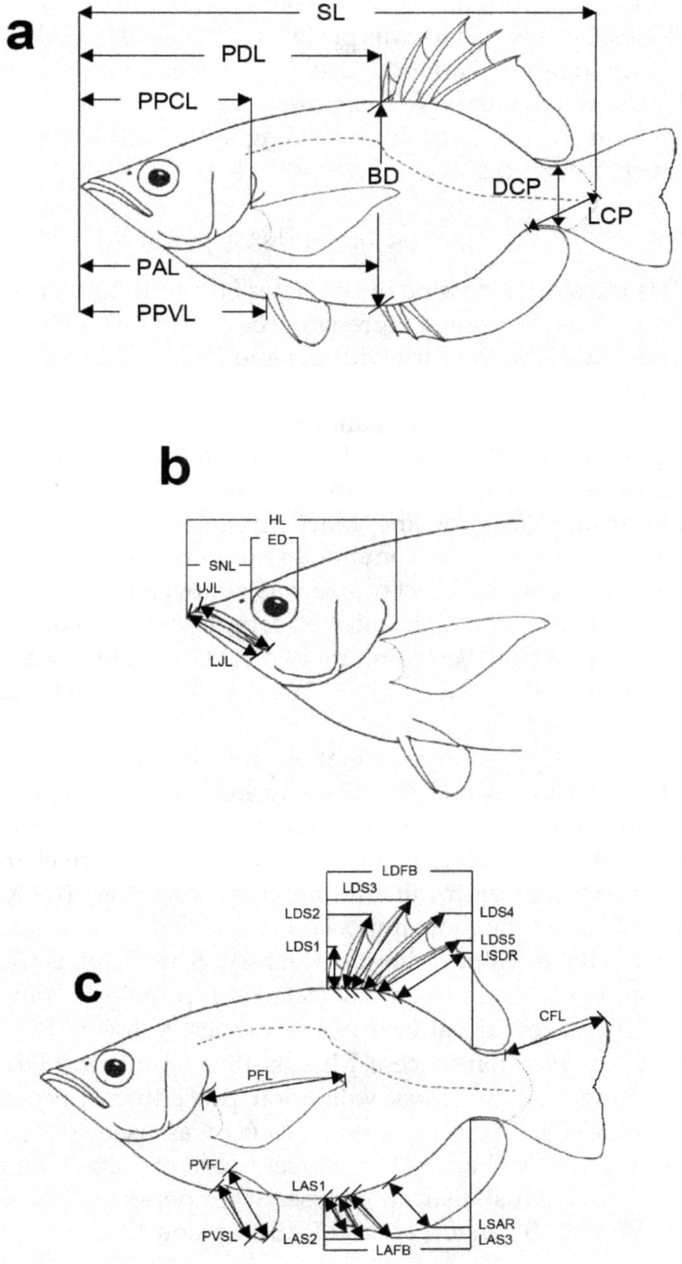

Figure 2.2. Morphometric measurements of archer fish (a) body measurements, (b) head measurements and (c) fin measurements.

Therefore, transformation of absolute measurements to size-independent shape characters was performed before the final analysis. In order to eliminate any variation resulting from allometric growth, all morphometric characters were therefore adjusted to an overall mean standard length ($\overline{SL}$) of 12.63 cm according to the following equation (Reimchen et al. 1985; Senar et al. 1994):

$$Y'_{ij} = \log Y_{ij} - b_j \left(\log SL_i - \log \overline{SL} \right)$$

where Y'_{ij} is the adjusted value of character j for individual i, Y_{ij} is the original value, b_j is the pooled regression coefficient of log Y on log SL, SL_i is the standard length of individual i and ($\overline{SL}$) is the overall mean standard length.

Adjusting to overall mean standard length removes size effects from morphometric data and has been shown to be an appropriate procedure for objective analysis of the data when there is size overlap among the groups (Claytor and MacCrimmon 1986). All statistical analyses were performed for combined sexes since all measurements were transformed and the effect of size was removed (Karakousis et al. 1993; Mamuris et al. 1998). The collected samples were predominantly made in both species (*T. jaculatrix* male n = 60, female n = 5; and *T. chatareus* male n = 60, and female n = 3) therefore sex effects were not considered in the present study.

The efficacy of size transformation was determined from the coefficient of determination (r^2) values of the log Y' *vs* log SL regression. A few of the morphometric characters could be considered as redundant, as the same part of the body is measured by two or more characters. Therefore, stepwise discriminant function analysis (DFA) was performed to extract the most important characters for differentiating species, using the F-value criterion (F-entry, 3.84; F-removal, 2.71). Selected characters were then subjected to principal component analyses (PCA) to reveal patterns of the species, followed by DFA to compute the classification success. Misclassification rates of DFA were calculated using holdout cross validation procedures proposed by Lachenbruch (1967). The Kappa statistics (κ) was used to determine the improvement over chance of the percent-correct classification rates (Titus et al. 1984). All statistical analyses were performed with SPSS version 15, MINITAB version 14, and PAST version 1.34.

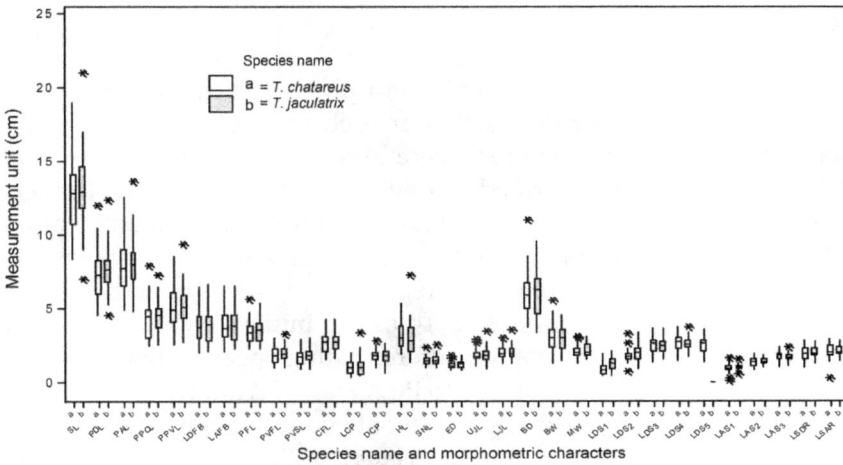

Figure 2.3. Box plot of morphometric characters of *T. chatareus* and *T. jaculatrix* (asterisk indicate outliers).

Table 2.2. Principal component loadings for the morphometric characters

Variable	PC1	PC2	PC3	PC4
PDL	-0.244	0.228	-0.245	-0.214
PPCL	-0.211	0.229	-0.138	-0.551
LDFB	-0.285	0.421	0.208	-0.051
PVFL	-0.461	0.062	0.134	0.144
DCP	-0.091	0.406	-0.270	0.136
HL	0.010	0.437	0.258	0.145
ED	-0.265	0.221	0.042	0.418
MH	-0.203	0.107	-0.097	-0.261
LDS1	-0.242	-0.245	-0.439	-0.039
LDS3	-0.114	0.005	-0.419	0.565
LAS1	-0.309	-0.306	0.112	0.099
LAS3	-0.178	-0.123	0.568	0.060
LSDR	-0.424	-0.310	-0.002	-0.043
LSAR	-0.321	-0.183	0.071	-0.093
Eigen value	3.191	2.475	1.5303	1.299
Proportion	0.228	0.177	0.1090	0.093
Cumulative	0.228	0.405	0.5140	0.607

The Outcome

Descriptive statistics for each of the morphometric characters are given in Figure 2.3. Twelve fish, six from each species were identified as extreme outliers and therefore were excluded from the analyses. The indication of the transformed characters to be free from the influence of size was provided by r^2 values. Prior to transformation almost 50% of the characters showed r^2 values above 0.50, whereas after transformation all the characters registered r^2 values of zero. Stepwise discriminant analysis identified 14 of the initial 31 characters as the most important characters for differentiating species; therefore, these characters were incorporated into PCA and DFA analyses.

The first four principal components cumulatively account 60.7% of the total morphological variation (Table 2.2). Almost all the loadings on PC1 (22.8%) are negative with no clear pattern. PC2 described 17.7% of the total variance, with length of dorsal fin base (LDFB), depth of

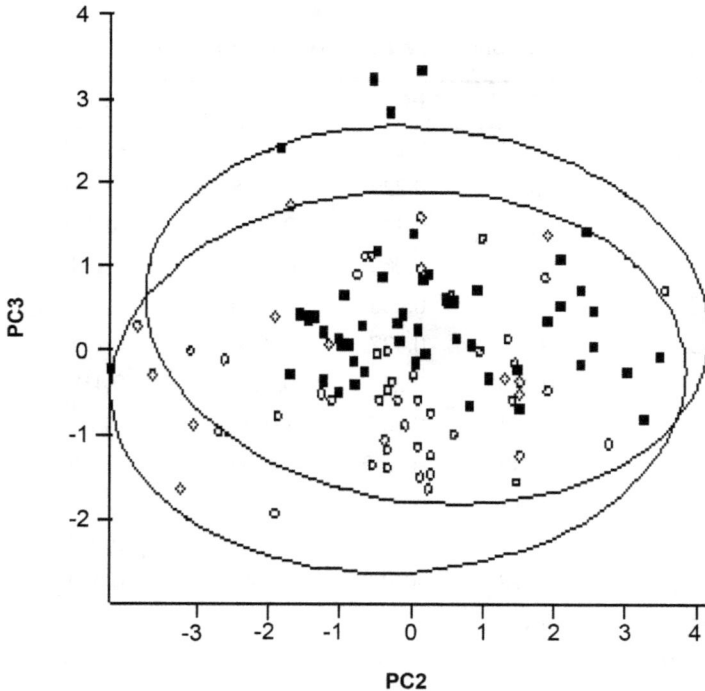

Figure 2.4. Scatter plot of PC2 and PC3 scores and 95% confidence ellipses of the scores for PCA using morphometric characters (*Toxotes chatareus* (solid square) and *Toxotes jaculatrix* (open circle).

Table 2.3. Standardised and unstandardised canonical discriminant function coefficients for morphometric characters

	Standardised function 1	Unstandardised function 1
PDL	0.332	10.884
PPCL	0.143	2.797
LDFB	-0.465	-8.496
PVFL	0.184	1.965
DCP	0.232	4.121
HL	-0.173	-2.169
ED	0.214	3.637
MH	0.250	4.288
LDS1	0.829	4.994
LDS3	-0.591	-8.601
LAS1	0.276	3.130
LAS3	-0.175	-2.590
LSDR	-0.438	-5.508
LSAR	0.132	1.121
Constant	-	-1.023

Table 2.4. Classification results of discriminant function analysis using morphometric characters of two archer fish species

		Species	Predicted group membership		
			T. chatareus	*T. jaculatrix*	Total
Original	Count	*T. chatareus*	47(87)	7(13)	54(100)
		T. jaculatrix	8(13.6)	51(86.4)	59(100)
Cross-validated	Count	*T. chatareus*	41(75.9)	13(24.1)	54(100)
		T. jaculatrix	14(23.7)	45(76.3)	59(100)

Figures in parentheses indicate percentage

caudal peduncle (DCP) and head length (HL) (large positive loadings) and anal spine length (LAS1) and length of soft dorsal ray (LSDR) (large negative loadings) being most highly correlated with PC2. PC3, which accounted for 10.9% of the total variance, contrasts the length of dorsal spines (LDS3 and LDS1; large negative loadings) and anal spine length (LAS3; large positive loading). The two species appear to differ on this PC, but with some overlap (Figure 2.4) with *T. chatareus* having on average shorter dorsal spine length and longer anal spine length than *T. jaculatrix*.

The discriminant function, tested using Wilks' lamda statistic, was significant ($\Lambda = 0.59$, $X^2_{14} = 54.8$, $P < 0.001$) indicating a relatively high degree of interspecies variance and that the means of the discriminant scores for the two species are different. The DFA picks out dorsal spine lengths (LDS1 and LDS3) as important discriminating characters, similar to the pattern seen on PC3 and to some extent the length of LDFB (Table 2.3).

The discriminant function managed to assign correctly 86.7% of the fish to the species and 76.1% after cross-validation, which is 73% (κ) better than what would have occurred by chance. The relatively high classification success provides support to the morphometric differences between the species (Table 2.4).

For the meristic characters, pelvic fin ray (PVFR), caudal fin ray (CFR), anal spine (AS), pelvic spine (PVS) and total vertebrae (TV) were constant between the species and were excluded in the analyses. The remaining meristic characters dorsal fin ray (DFR), pectoral fin ray (PCFR), anal fin ray (AFR), lateral line scale (LLS) were subjected to PCA analyses. Univariate comparisons of these characters between species were significant ($P < 0.05$) except for dorsal fin ray (DFR) (Tables 2.5, 2.6 and 2.7).

Table 2.5. Univariate comparisons of meristic characters of *T. chatareus* and *T. jaculatrix*

	T. chatareus		T. jaculatrix	
	Mean	**S.E.**	**Mean**	**S.E.**
DFR	11.96[a]	0.07	11.88[a]	0.07
PCFR	12.76[a]	0.06	14.49[b]	0.09
AFR	16.19[a]	0.07	15.81[b]	0.09
LLS	32.50[a]	0.45	28.73[b]	0.13

[a,b] Different letters indicates significant difference at $P \leq 0.05$

Three PCs were extracted accounting for 94% of the total variation. PC1 accounted for 52.5% and the loadings were all positive thus is not

Table 2.6. Principal component loadings for the meristic characters

Variable	PC1	PC2	PC3	PC4
DFR	0.639	-0.126	0.114	-0.750
PCFR	0.201	-0.794	0.437	0.371
AFR	0.608	-0.022	-0.675	0.419
LLS	0.427	0.594	0.584	0.353
Eigen value	2.099	1.279	0.412	0.209
Proportion	0.525	0.320	0.103	0.052
Cumulative	0.525	0.845	0.948	1.000

Table 2.7. Standardised and unstandardised canonical discriminant function coefficients for meristic characters

	Standardised function 1		Unstandardised function 1
DFR	0.296	DFR	0.495
PCFR	-1.441	PCFR	-2.250
AFR	0.757	AFR	1.066
LLS	0.431	LLS	0.162
		(Constant)	2.792

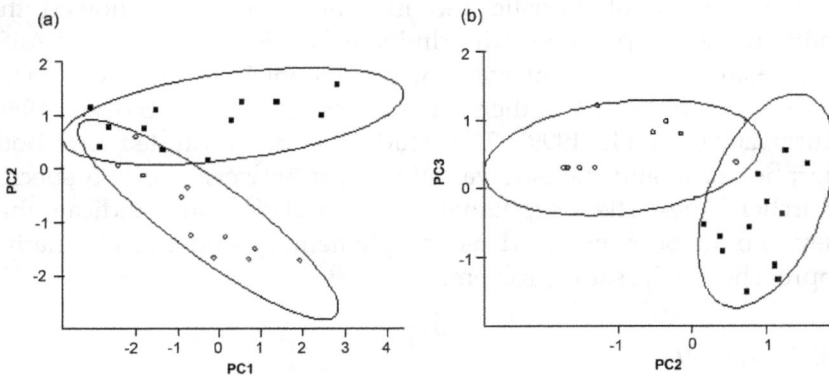

Figure 2.5. Scatter plot (a) PC1 and PC2, (b) PC2 and PC3 scores and 95% confidence ellipse of the scores for PCA using meristic characters (*T. chatareus* (solid square) and *T. jaculatrix* (open circle)).

particularly informative, probably describing general size axis. The loadings on PC2 (32%) and PC3 (10.3%) are both positive and negative. PC2 contrasts between PCFR (negative loading) and LLS (positive loading) (Table 2.6). Bivariate plot of PC1 and PC2 scores revealed that separation of *T. chatareus* and *T. jaculatrix* is evident on PC2. *T. chatareus* have positive scores while *T. jaculatrix* have negative scores, albeit with slight overlap (Figure 2.5).

The squared canonical correlation of the discriminant function was 0.86, suggesting that a high proportion of the total variance of meristics was attributable to differences between species (Table 2.7). Discriminant analyses managed to correctly classify 100% of the samples.

Morphometric character analysis demonstrated that although the two species are less distinct, 76.1% of fish were ascribed to the correct species cluster. The dorsal spine lengths (LDS1 and LDS3) and anal spine length (LAS3) and to some extent, length of dorsal fin base (LDFB) were found to be important discriminating morphometric characters in the present study with *T. chatareus* in general having shorter dorsal spine length and longer anal spine length than *T. jaculatrix*.

Analyses of meristic characters revealed that PC2 and the discriminant function are highly correlated with pectoral fin rays (negative loading) and lateral line scales (positive loading) and anal fin rays (positive). *T. chatareus* can be distinguished from *T. jaculatrix* by having greater number of lateral line scales, lower pectoral fin ray counts and a slightly higher anal fin ray counts.

Comparison of meristic and morphometric traits showed the ability of the morphometric discriminant function to correctly classify individuals, in agreement with the results obtained by discriminant function analyses with other fish species (Meng and Stocker 1984; Murta 2000; Tudela 1999). This study has demonstrated that both morphometric and meristic variation exist between the two species of archer fishes. The morphometric and meristic results indicate that they should be considered as complementary and not alternative approaches to the same problem.

Conclusion

T. chatareus and *T. jaculatrix* are aggressive in nature and they sometimes lose their fins and spines because of their hostile behaviour. Moreover, body colouration, which was previously used as one of the identifying characters of these two species usually depends on the habitat. The present study has uncovered some morphological (i.e. morphometric and meristic) distinctions between the two closely related archer fishes using multivariate techniques. This study demonstrates that *T. chatareus* and *T. jaculatrix* from Sungai Santi, Johor coastal waters were different from one another in both morphometric and meristic traits. The best statistical classifications of these groups using multivariate discriminant analyses were obtained using meristic characters, while morphometric characters provided comparatively less evidence of differentiation. Therefore, this chapter can be used to identify these two species more precisely as colour can be regarded as a plastic character (most variable characteristics). There could also be a possible link between the observed meristic variability with differences in habitat, and prey-predatory relationship. However, the true reasons for the observed meristic and morphometric variability should be studied further using appropriate sampling design that includes different localities as well as predatory behaviour of the fishes.

Age Composition and Growth of Archer Fish

Population growth of fishes can be described in two ways, first by describing size (length) at age by the Ludwig von Bertalanffy growth function and secondly, by length weight relationship (LWR) analysis. LWR is a useful tool in fishery assessment, which helps in predicting weight from length acquired in yield assessment and in the calculation of the standing crop biomass. LWR also allows fish condition to be estimated. Condition factor is a quantitative parameter of the state of well-being of the fish that will determine present and future population success by its influence on growth, reproduction and survival. The condition of a fish reflects recent physical and biological circumstances, and fluctuates by interaction among feeding conditions, parasitic infections and physiological factors.

To date there is no information on age, and growth of *Toxotes chatareus* and *Toxotes jaculatrix* or any toxotid, in Malaysia or elsewhere. Consequently, lack of adequate knowledge on age and growth of *T. chatareus* and *T. jaculatrix* remains an impediment to the definition of sound management strategies for these fascinating fishes. Therefore, the aim of this chapter is to provide information on age, length-weight relationship, condition factors and individual growth rate of *T. chatareus* and *T. jaculatrix*.

The Approach

Sampling was carried out monthly in Sungai Santi, Johor coastal waters, Malaysia from July 2007 to June 2008. A total of 350 fish (*T. chatareus*, n = 145; *T. jaculatrix*, n = 205) were collected using cast, three layered trammel nets, hand line (rod fishing) and cages (Chapter 3). Several individuals (n = 164, 82 *T. chatareus* and 82 *T. jaculatrix*) of varying sizes were saved for ageing study.

Age was determined by using scales and otolith of 164 fishes. For each specimen about 10 scales were sampled from the central portion of the body below the lateral line (Chung and Woo 1999; Elp and Sen 2009; Paul 1967). The scales were treated in 0.5% ammonia solution for at least 2 days, rinsed thrice with distilled water, dried and mounted between two microscope slides. Regenerated scales were discarded. On the other hand, sagittae otoliths were taken by dissection of the dorsal part of the fish head, where upper head sections were cut diagonally just below the base of the periscopic eyes. The sagittae otolith were processed to the desirable thickness and mounted on an epoxy-resin block prior to grinding using fine carbonised sand papers (# 1000, # 3000) (Mazlan and Rohaya 2008; Secor et al. 1991). There was no disparity observed between sizes and shapes of right and left otoliths of same individuals, consequently right otoliths were used for ageing of fishes (Figure 3.1). The mounted scales and otoliths were labeled,

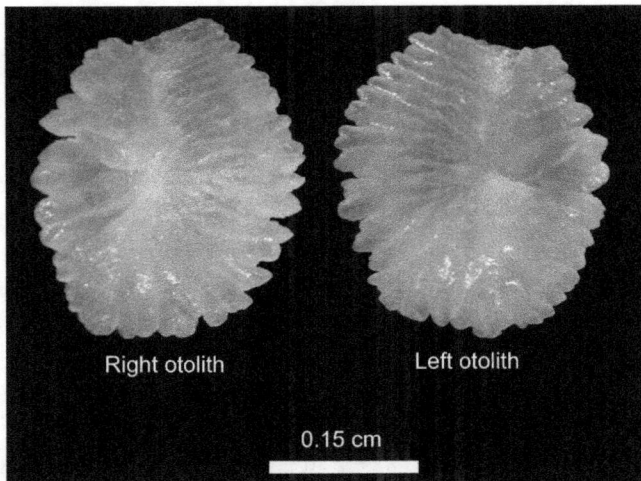

Right otolith   Left otolith

0.15 cm

Figure 3.1. External morphology of right and left sagittae otolith of archer fish.

observed and photographed under a HITACHI Table Top Scanning Microscope TM-1000 and ZEISS Stereo Microscope Stemi DV4/DR. The daily increment or rings (age in day) and annulus (age in year) of the scale as well as annulus of otolith were repeatedly counted using the Adobe™ graphic software (Mazlan and Rohaya 2008) to determine the ages of two archer fishes.

Population growth of *T. chatareus* and *T. jaculatrix* was estimated based on the LWR analysis. The relationship between the length and weight of a fish is usually expressed by the equation $W = aL^b$ (Abdallah 2002; Ricker 1973), where W is body weight (g), L is total length (cm), a is the intercept and b is the slope (fish growth rate) (Beverton and Holt 1996). Determination of a and b values were done using a non-linear regression for which curve fitting was carried out by a non-linear iterative method using Levenberg-Marquardt and Simplex algorithms for obtaining best convergence X^2 goodness of fit values using a computer programme (Microcal Origin™ Version 6.0).

The degree of adjustment of the model studied was assessed by the coefficient of determination (r^2). Student's *t*-test was applied to verify whether the declivity of regression (constant b) presented a significant difference of 3.0, indicating the type of growth: isometric ($b = 3.0$), positive allometric ($b > 3.0$) or negative allometric ($b < 3.0$) (Spiegel 1991) using MINITAB version 14 software. In all cases a statistic significance of 5% was adopted. The Fulton condition factor (K) was calculated from the expression (Bauchot and Bauchot 1978): $K = 1000 \ W/L^3$, where W is the whole-body weight in gram (g) and L the total length in centimeter (cm). The relative condition factor (K_n) was calculated according to Godinho (1997): $K_n = W/aL^b$. The parameters a and b of the LWR were used to calculate the relative condition factor (K_n).

A single von Bertalanffy growth curve was fitted to the lengths at age of the individuals of each species, using a non-linear regression for which curve fitting was carried out by a non-linear iterative method using Levenberg-Marquardt and Simplex algorithms for obtaining best convergence X^2 goodness of fit values using a computer programme, Microcal Origin™ Version 6.0. The von Bertalanffy equation is $L_t = L_\infty (1 - \exp^{-K(t-t_0)})$, where L_t is the total length (cm) at age t (years), L_∞ is the predicted asymptotic length (cm) predicted by the equation, K is the growth coefficient (year^{-1}) and t_0 is the hypothetical age (years) at which fish would have zero length, if growth followed as predicted by the equation (von Bertalanffy 1938; Ricker 1975; Punt and Hughes

1992). The growth performance index was obtained from the equation: ϕ' (phi prime) $= \ln K + 2^* \ln L_\infty$ (Munro and Pauly 1983; Pauly and Munro 1984).

The Outcomes

Estimated ages in the scales (daily increments and annulus analysis, Figures 3.2a, c) and otoliths (annulus analysis, Figure 3.2b) ranged

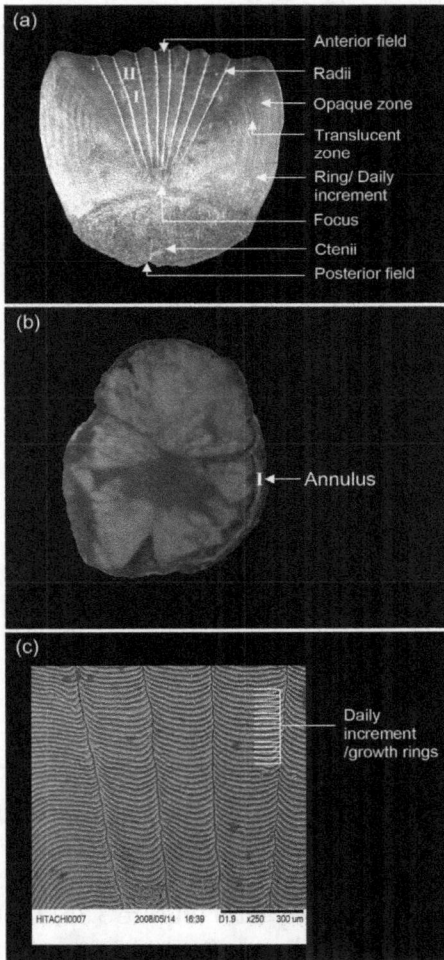

Figure 3.2. (a) Scale from a 2+ year-old individual and (b) sagittae otolith from a 1+ year-old individual of an archer fish, (c) electron micrograph displaying daily increment/growth rings on the scale.

from 245-1540 days or (0+ to 4+ years) (Tables 3.1-3.4). The results demonstrated for *T. chatareus* males in the size ranged from 8.5 to 20.0 cm TL, the estimated ages were from 245-989 days (0+ to 2+ years), while for females, the size ranged from 9.8 to 22.5 cm TL, the estimated ages were from 342-1523 days (0+ to 4+ years), respectively (Figure 3.3a, Tables 3.1 and 3.2).

Table 3.1. Age-length key for *T. chatareus* males collected from Sungai Santi, Johor coastal waters, Malaysia

Length (TL, cm)	0+	1+	2+	3+	4+
8.5-9.5	6				
9.5-10.5	5				
10.5-11.5	10				
11.5-12.5	2	8			
12.5-13.5		8			
13.5-14.5		6			
14.5-15.5		4			
15.5-16.5					
16.5-17.5		1			
16.5-17.5			1		
17.5-18.5			1		
18.5-19.5			1		
19.5-20.5					
20.5-21.5					
21.5-22.5					
22.5-23.5					
Total (n)	23	27	3		

T. jaculatrix males ranged in size from 8.5 to 19.0 cm in TL for fishes with estimated ages of 246-929 days (0+ to 2+ years), while for females, size ranged from 8.7 to 23.0 cm TL for fishes with estimated ages of 283-1540 days (0+ to 4+ years) (Figure 3.3b, Tables 3.3, 3.4). The dominant age classes recorded were 0+ and 1+ years in males of both and 1+ and 2+ years in females of both species. The results indicated that females exhibited a lower growth rate ($b = 1.82, 1.97$) than males ($b = 1.79, 1.72$) in *T. chatareus* and *T. jaculatrix*, respectively (b measures changes in age with length; consequently larger *b* values means lower growth rate and vice versa) (Figure 3.3). The results also showed that the number of daily ring counts was significantly correlated with fish size (r^2 = 0.98, 0.99 for females; r^2 = 0.99, 0.98 for males ($P < 0.05$), *T. chatareus* and *T. jaculatrix*, respectively).

Table 3.2. Age-length key for *T. chatareus* females collected from Sungai Santi, Johor coastal waters, Malaysia

Length (TL, cm)	0+	1+	2+	3+	4+
8.5-9.5					
9.5-10.5	2				
10.5-11.5		3			
11.5-12.5		4			
12.5-13.5		4			
13.5-14.5		6			
14.5-15.5		2			
15.5-16.5		1	4		
16.5-17.5			1		
16.5-17.5			1		
17.5-18.5					
18.5-19.5					
19.5-20.5					
20.5-21.5					
21.5-22.5					1
22.5-23.5					
Total (n)	2	20	6		1

Length-weight relationship was derived from 145 *T. chatareus* and 205 *T. jaculatrix* samples. The LWR for combined sexes, males and females of *T. chatareus* and *T. jaculatrix* is shown in Figures 3.4 and 3.5. The intercept *a* for combined sexes of *T. chatareus* and *T. jaculatrix* were 0.007 and 0.012, and coefficient of determination r^2 were 0.971 and 0.959 respectively (Figure 3.4). Intercept *a* for male *T. chatareus* was 0.01 and female was 0.005 and coefficient of determination r^2 for male was 0.969 and female was 0.970, while in *T. jaculatrix* intercept *a* was 0.015 for male and 0.007 for female and coefficient of determination r^2 was 0.958 for male and 0.950 for female (Figure 3.5). The values of the

Table 3.3. Age-length key for *T. jaculatrix* males collected from
Sungai Santi, Johor coastal waters, Malaysia

Length (TL, cm)	0+	1+	2+	3+	4+
8.5-9.5	5				
9.5-10.5	5				
10.5-11.5	8				
11.5-12.5	5	6			
12.5-13.5		6			
13.5-14.5		7			
14.5-15.5		3			
15.5-16.5		2			
16.5-17.5		3			
16.5-17.5			2		
17.5-18.5			1		
18.5-19.5					
19.5-20.5					
20.5-21.5					
21.5-22.5					
22.5-23.5					
Total (n)	23	27	3		

slope or exponent b for males (3.246), females (3.465), and combined sexes (3.353) were significantly ($P < 0.05$) higher than 3, exhibiting a positive allometric growth for *T. chatareus*, whereas the exponent b value for male (3.076) was close to 3.0 ($P > 0.05$), showing isometric growth pattern (i.e., changing the body form following the cube law (volume $= L^3$) for *T. jaculatrix*. In contrast, the estimated b values for female (3.310) and combined sexes (3.160) were significantly ($P < 0.05$) higher than 3 and therefore, exhibited a positive allometric growth for *T. jaculatrix*, indicating that weight increases faster than length (Figures 3.4 and 3.5).

The mean monthly Fulton condition factor (K) and relative condition factors (K_n) of male *T. chatareus* reached their maximum in later months (i.e., July, through September). The mean monthly K, and K_n for male *T. chatareus* rose from 15.65, 1.29 in July to reach a peak of 17.37, 1.63 in September respectively and then declined precipitously

Table 3.4. Age-length key for *T. jaculatrix* females collected from Sungai Santi, Johor coastal waters, Malaysia

Length (TL, cm)	0+	1+	2+	3+	4+
8.5-9.5	1				
9.5-10.5	1				
10.5-11.5	1	1			
11.5-12.5		3			
12.5-13.5					
13.5-14.5		3			
14.5-15.5		4			
15.5-16.5		2	2		
16.5-17.5			5		
16.5-17.5			4		
17.5-18.5			1		
18.5-19.5					
19.5-20.5					
20.5-21.5					
21.5-22.5					
22.5-23.5					1
Total (n)	3	13	12		1

to 15.98, 1.47 in October and reached their lowest mean K and K_n values 14.21, 0.95 in November (Figures 3.6).

The mean monthly K and K_n of male *T. jaculatrix* followed a similar trend rising from 15.69, 1.45 in July to a maximum of 17.43, 1.68 in September respectively (Figure 3.7). The mean monthly K and K_n of female *T. chatareus* and *T. jaculatrix* were higher than those of the males (Figures 3.6 and 3.7). The mean monthly K and K_n of female *T. chatareus* increased from 16.04, 1.42 in July to 17.96, 1.79 in September and then remained at 16.56, 1.6 in October respectively, before declining from November to December (Figure 3.6). The mean monthly K and K_n of female *T. jaculatrix* showed a similar trend, increasing from 16.98, 1.52 in July to 18.85, 1.83 in September and then remained 17.13, 1.54 in October, before declining from November to December (Figure 3.7).

The relationship between age and length was adequately described by the von Bertalanffy growth curves (Figure 3.8 and 3.9). The

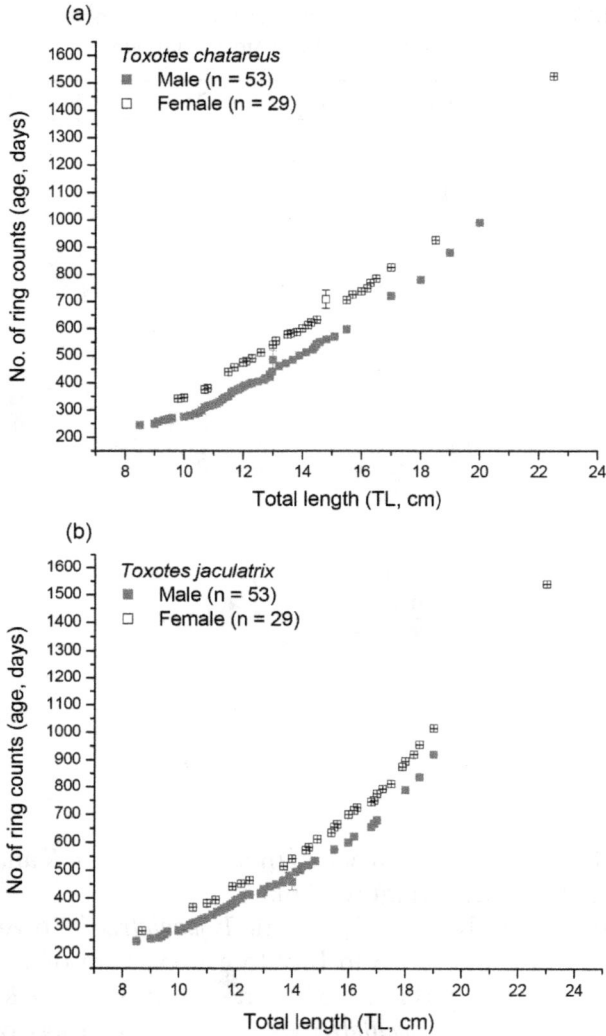

Figure 3.3. Relationship between number of daily ring counts (from scale, age in days) and total length (cm) of (a) *T. chatareus* and (b) *T. jaculatrix* (females r^2 = 0.98, 0.99, males r^2 = 0.99, 0.98, P < 0.05, *Toxotes chatareus* and *Toxotes jaculatrix*, respectively).

estimated von Bertalanffy growth parameters were: L_∞ = 26.12, 23.52 cm for males and 28.57, 25.87 cm for females, K = 0.41, 0.66 year^{-1} for males and 0.35, 0.47 year^{-1} for females, and t_0 = –0.26, –0.03 for males and –0.23, –0.10 for females of *T. chatareus* and *T. jaculatrix*, respectively

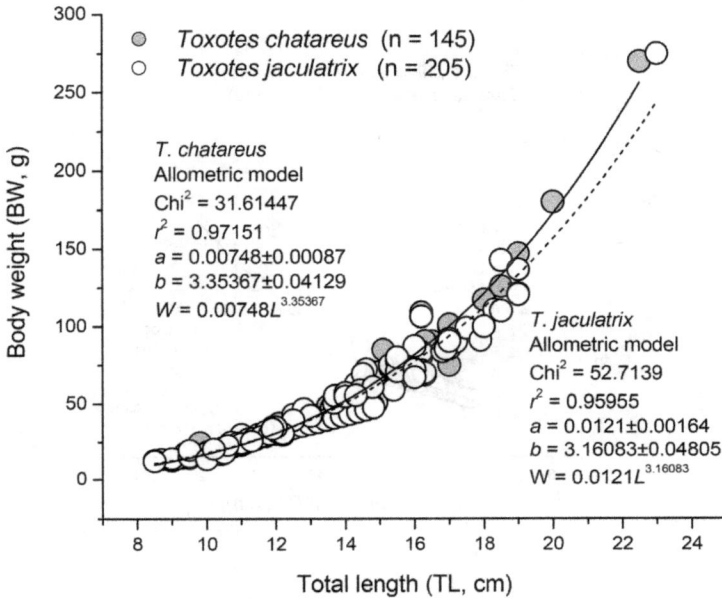

Figure 3.4. Length-weight relationships of *T. chatareus* and *T. jaculatrix* (combined sex) (solid regression line represents non-linear fit of *T. chatareus* and dashed regression line represents non-linear fit of *T. jaculatrix*).

(Table 3.5, Figures 3.8 and 3.9). The estimated growth performance indexes or phi-prime: (ϕ') for males were 2.45, 2.56 and for females were 2.46, 2.49 in *T. chatareus* and *T. jaculatrix*, respectively.

Ageing of fishes from tropical waters has been reported through annual increments in calcified structures such as scales (Chung and

Table 3.5. Parameters for the von Bertalanffy growth curves and related statistics fitted to the lengths at age of *Toxotes chatareus* and *Toxotes jaculatrix* in Sungai Santi, Johor coastal waters, Malaysia

Species	Sex	$L_\infty \pm$ S.E.	$K \pm$ S.E.	$t_0 \pm$ S.E.	r^2	X^2
Toxotes	Male	26.12±1.31	0.41±0.04	-0.26±0.04	0.99	0.04
chatareus	Female	28.57±1.00	0.35±0.02	-0.23±0.05	0.99	0.05
Toxotes	Male	23.52±0.48	0.66±0.03	-0.03±0.02	0.99	0.02
jaculatrix	Female	25.87±0.57	0.47±0.02	-0.10±0.04	0.99	0.04

L_∞ = Asymptotic length, K = Growth coefficient and t_0 = Hypothetical age at which fish would have zero length, S.E. = Standard error; r^2 = Coefficient of determination, X^2 = chi-square

(a)

Female
Allometric model
$Chi^2 = 62.27186$
$r^2 = 0.96979$
$a = 0.00544 \pm 0.00142$
$b = 3.46561 \pm 0.09019$
$W = 0.00544L^{3.46561}$

Male
Allometric model
$Chi^2 = 21.01694$
$r^2 = 0.96905$
$a = 0.01 \pm 0.00132$
$b = 3.24615 \pm 0.04826$
$W = 0.01L^{3.24615}$

○ *Toxotes chatareus* (female, n = 35)
○ *Toxotes chatareus* (male, n = 110)

(b)

Female
Allometric model
$Chi^2 = 71.23774$
$r^2 = 0.95098$
$a = 0.00782 \pm 0.00195$
$b = 3.31054 \pm 0.087$
$W = 0.00782L^{3.31054}$

Male
Allometric model
$Chi^2 = 42.94132$
$r^2 = 0.95844$
$a = 0.01542 \pm 0.00248$
$b = 3.07657 \pm 0.05774$
$W = 0.01542L^{3.07657}$

○ *Toxotes jaculatrix* (female, n = 63)
○ *Toxotes jaculatrix* (male, n = 142)

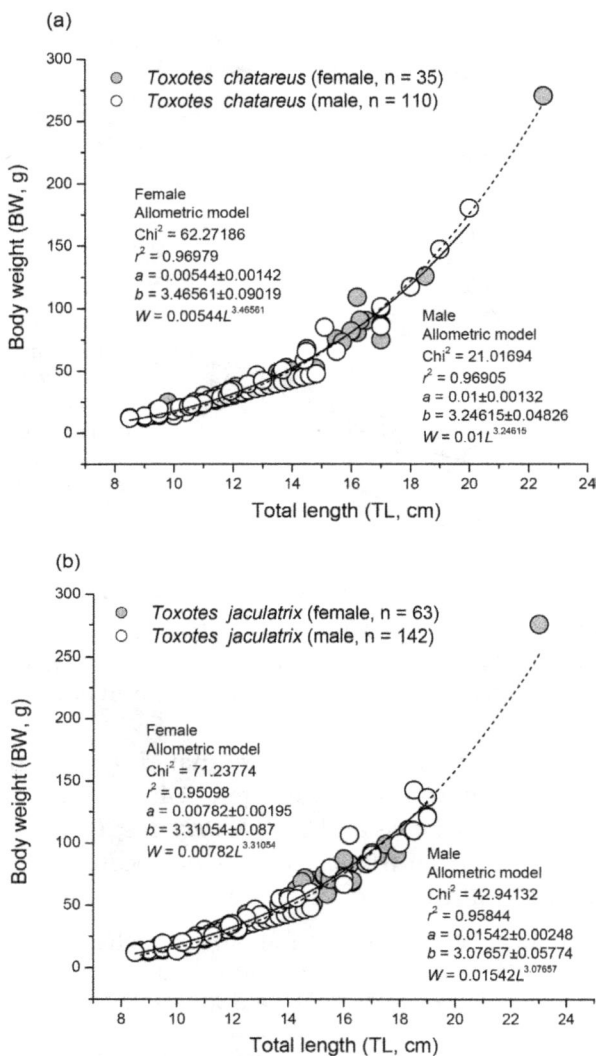

Figure 3.5. Length-weight relationships between sexes (a) *T. chatareus* and (b) *T. jaculatrix* (dashed regression line represent non-linear fit of female specimens and solid regression line represent non-linear fit of male specimens in both species).

Woo 1999; Mayekiso and Hecht 1988; Werder and Soares 1985), dorsal and pectoral spines (Ezenwa and Ikusemiju 1981; Pantulu 1961), vertebral centra (Brown and Gruber 1988), and otoliths (Fowler and Doherty 1992; Mazlan and Rohaya 2008). Scales are the easiest to

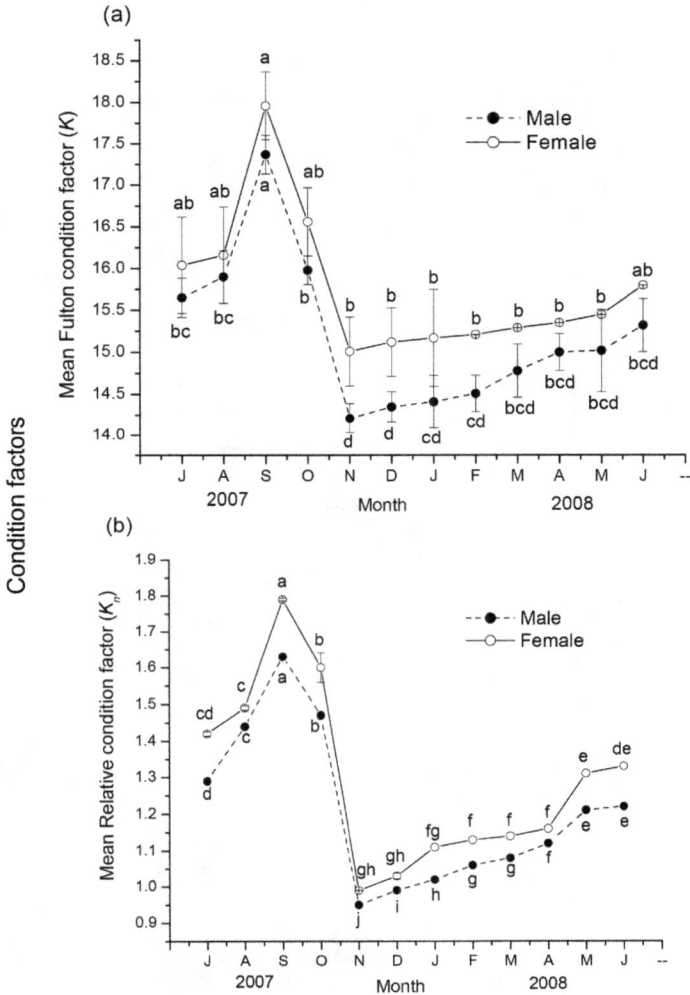

Figure 3.6. Mean monthly condition factors of *T. chatareus* (a) Fulton condition factors (K), (b) relative condition factors (K_n). Different letters above the mean values indicate significant differences of mean condition factors at $P < 0.05$.

collect and process. Using scales as structures for ageing also avoids sacrificing the specimens like in ageing methods employing otoliths. Moreover, otolith require prolonged preparation than removal of scale, and otolith analysis demands special equipment for sectioning in addition to being labour intensive (Sullivan et al. 2003).

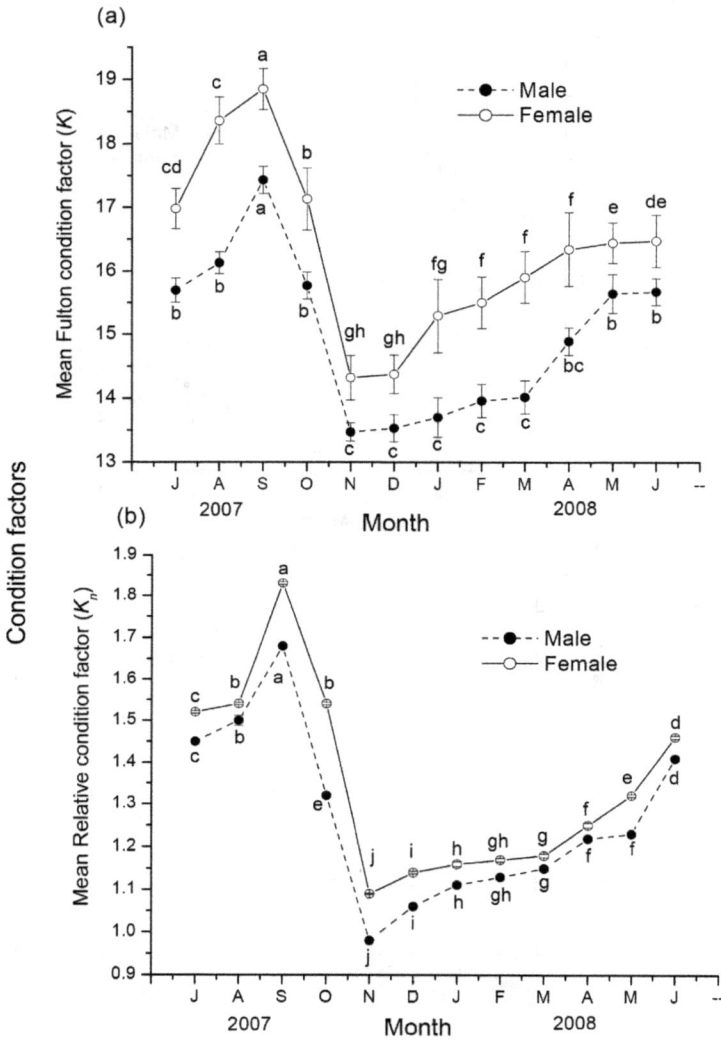

Figure 3.7. Mean monthly condition factors of *Toxotes jaculatrix* (a) Fulton condition factors (K), (b) relative condition factors (K_n). Different letters above the mean values indicate significant differences of mean condition factors at $P < 0.05$.

Besides, it was also noticed in the present study that otoliths of the archer fishes often proved to be useless for age determination due to high opacity and being delicate to handle. Consequently, daily increments or growth rings were not obtained from the otolith hard structure. Using scales for fish ageing, however, suffers drawbacks

(a)

$Toxotes\ chatareus$ (male, n = 53)
von Bertalanffy growth curve:

$L_t = 26.12\,(1 - exp^{-0.41(t+0.26)})$

(b)

$Toxotes\ chatareus$ (female, n = 29)
von Bertalanffy growth curve:

$L_t = 28.57\,(1 - exp^{-0.35(t+0.23)})$

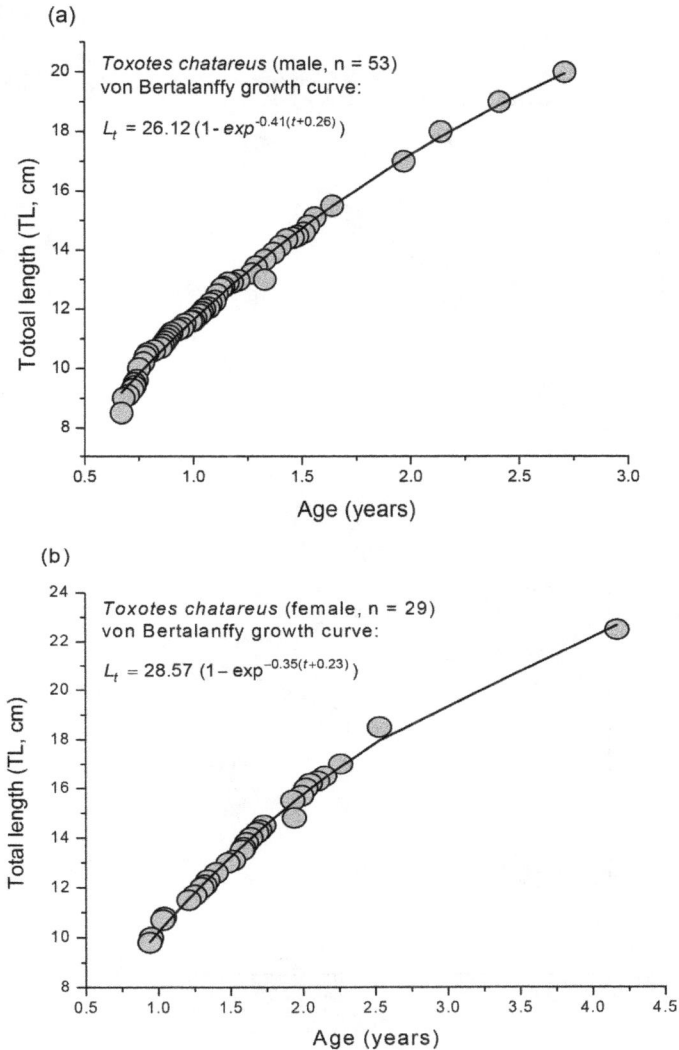

Figure 3.8. von Bertalanffy growth curves fitted to the length at age of
T. chatareus (a) males and (b) females collected from Sungai Santi, Johor
coastal waters, Malaysia. Sample sizes (*n*) shown on each figure.

like difficulties in reading annuli, low precision (Lowerre-Barbieri et
al. 1994), and that scale ages may become inaccurate when growth
becomes asymptotic (Beamish and McFarlane 1987, Shepherd 1988).
In this research, the reliability of scale readings was increased by
sampling scales only from a fixed position, where the scales have large

(a)

(b)

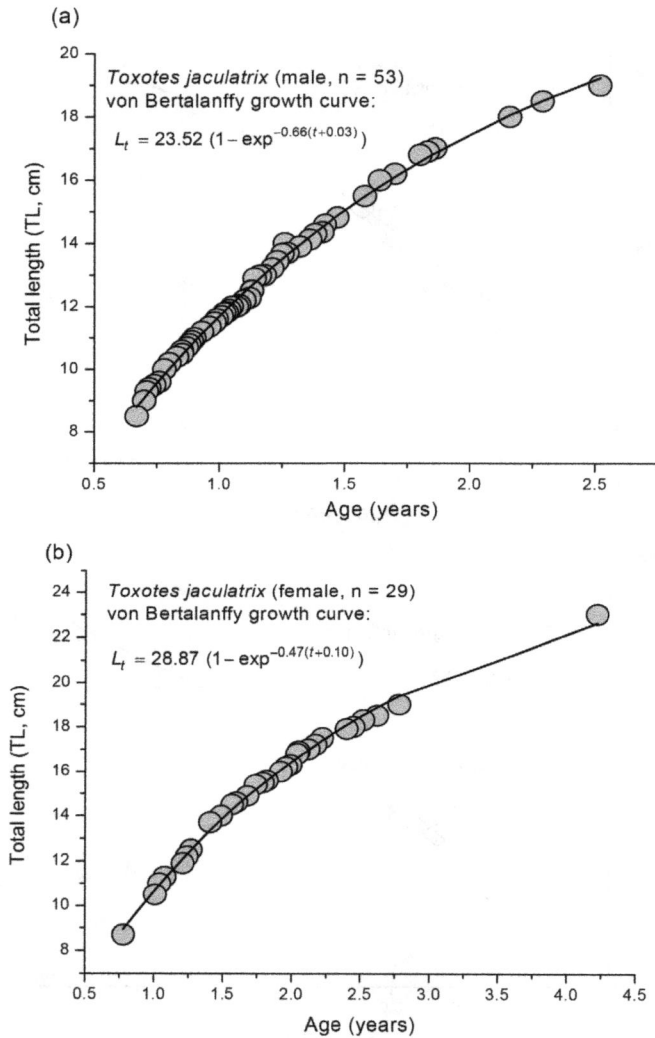

Figure 3.9. von Bertalanffy growth curves fitted to the length at age of *T. jaculatrix* (a) males and (b) females collected form Sungai Santi, Johor coastal waters, Malaysia. Sample sizes (*n*) shown on each figure.

uniform size, better symmetry, and high legibility. On the other hand, as extremely senescent specimens were unavailable in this research, annuli readings have been relatively legible and reliable. Moreover, the estimated ages from scales daily increment and annulus analysis were further validated with otolith (Figure 3.2).

In the current study, it was observed that *T. chatareus* and *T. jaculatrix* (except male samples of *T. jaculatrix*) exhibit similar growth patterns (i.e., positive allometry: fish becomes more rotund as length increases). The slopes or exponents *b* for male, female, and both sexes for *T. chatareus* were significantly ($P < 0.05$) higher than 3.0, reflecting a positive allometric growth, whereas in *T. jaculatrix* the values of exponent *b* for males were close to 3.0 ($P > 0.05$), showing isometric growth pattern. This isometric growth pattern is probably attributed to sex. The estimates of the parameter *b*, varying between 2 and 4 (Bagenal and Tesch 1978), remain within the expected range (3.076-3.465), with a mean *b* value of 3.268 ($\pm$ 0.056) for all the species. TL and BW of both species are highly correlated ($r^2 = 0.95$-0.96). The length weight relationships of *T. chatareus* and *T. jaculatrix* have not been previously recorded in Malaysia or elsewhere, preventing the comparison with previous results. However, for more precise weight estimations, the application of these length-weight relationships should be limited to the observed length ranges; otherwise, it may be erroneous (Froese 1998; Petrakis and Stergiou 1995).

Even though the change of *b* value depends primarily on the shape and fatness (size) of the species, various factors may be responsible for the differences in parameters of the length-weight relationships among seasons and years, including temperature, salinity, food (quantity, quality, and size), sex, time of year, and stage of maturity (Bagenal and Tesch 1978; Gonçalves et al. 1997; Özaydin and Taskavak 2007; Pauly 1984; Sparre 1992; Taskavak and Bilecenoglu 2001). However, none of them were taken into consideration in the present study.

It was observed in the present study, that Fulton condition factors (*K*), for all species were of values 1 and above which indicates that fish species are doing well in the study area. When *K* is greater than unity, the fish species is heavy. Bagenal and Tesch (1978) documented that for mature fresh and brackish water fish, condition factor aught to be in the range of 2.9-4.8. The *K* values in the present study were relatively high for the reason that a more homogenous formula of condition factor ($K = 100 \, W/L^3$) was used (Bouchot and Bouchot 1978). However, Safran (1992) stated that the parameters *a* (condition factor) and *K* were judged to be less important in comparative studies, since these parameters were closely correlated with *b*. As a matter of fact, for applied ichthyological studies, only *b* seems to be important as a key parameter in estimating population growth through LWR (Kimmerer et al. 2005; Safran 1992).

The average K and K_n of both *T. chatareus* and *T. jaculatrix* were lowest in November whilst highest in September for both sexes; it is the time when gonads of the most individual fish were fully grown thus contributing to the higher mean K and K_n values (Figures 3.6 and 3.7). In this study it seems that K and K_n have a correlation with the monthly changes in maturity (Gonadosomatic index of fish), suggesting on the one hand that in November both species may start the reproductive period in the Johor coastal waters and, on the other hand in September, fish may already recover.

The K and K_n values can be influenced by certain extrinsic factors namely, changes in temperature and photoperiod (Youson et al. 1993). For the *T. chatareus* and *T. jaculatrix* in the study areas, the temperature and photoperiod elements might not be significant factors because Malaysia in general experiences no great difference on those parameters throughout the year compared to temperate countries (Samat et al. 2008).

The von Bertalanffy growth parameters calculated in this study indicated that *T. chatareus* is a slow growing species with $K = 0.41$ year^{-1} for males and 0.35 year^{-1} for females unlike the *T. jaculatrix* with $K = 0.66$ year^{-1} for males and 0.47 year^{-1} for females. However, the estimated growth parameter K in the present study was higher than reported by Zorica et al. (2005), who obtained $K = 0.192$, year^{-1}, and lower than reported by Mazlan and Rohaya, who obtained $K = 1.4$ year^{-1} for other tropical fish species. Therefore, the estimation of the phi-prime indices (ϕ') from our study was somewhat higher ($\phi' = 2.49 \pm 0.02$) than the estimated value ($\phi' = 2.25$) obtained by Zorica (2005) and lower than the estimated value ($\phi' = 3.1$) obtained by Mazlan and Rohaya (2008). These variations may be related to habitat characteristics, physico-chemical parameters of the environment or abundance and accessibility of prey items (Bilecenoglu 2009; Morales-Nin and Ralston 1990).

Conclusion

Commonly practiced fish population demographic research (namely estimate asymptotic length, growth coefficient etc.) uses larger amount of samples whilst the present study used a total of 164 specimens. Therefore, estimates of the von Bertalanffy growth parameters provide only preliminary information on the growth patterns of *T. chatareus* and *T. jaculatrix* from the limited study area. However, a high degree of positive correlation between TL and ages of both species is indicated by high values of coefficient of determination r^2. The theoretical asymptotic length for both *T. jaculatrix* and *T. chatareus* was bigger (30 cm and 50 cm TL respectively (Allen 1978, 2001)) than the maximum size of fishes recorded in the present study (22.5 cm and 23.0 cm TL, respectively). This chapter has provided the basic information on the age, growth and conditions of *T. chatareus* and *T. jaculatrix* that would be useful for fishery biologists, managers and conservationist communities to impose adequate regulations for sustainable management of these fascinating fishes in the Johor coastal waters and nearby areas of Malaysia. Moreover, the study also suggests that, scale can be used as a key hard or calcified structure to estimate ages without scarifying these rare and attractive fishes.

Feeding Techniques and Stomach Content Analysis of Archer Fish

The archer fishes are a family of euryhaline fish that have the ability to shoot down aerial insects with a precisely aimed jet of water as their feeding technique. Since their eyes remains entirely below the surface during positioning and shooting, the fishes have a potentially serious optical problem which they must correct, i.e., refraction at the air-water interface. *Toxotes chatareus* and *Toxotes jaculatrix,* the two species of archer fishes are often found in large schools in the natural environment and that the spitter does not always get the prey because of the aggression amongst them (Personal observation; Das's unpublished data). However, no comprehensive study has been conducted on the spitting success rate and accuracy of archer fishes' relative to group size. Previously everyone else works on one or the other and treats them as equals but in fact they have not been compared directly.

The quality and quantity of food are among the most important exogenous factors directly affecting growth and, indirectly, maturation and mortality in fish, thus being ultimately related to fitness (Wootton 1990). Traditionally, information on the quality and quantity of food consumed by fish, can be derived from their feeding studies (Jennings et al. 2001).

Despite the magnitude of feeding studies for fisheries and ecosystem research, relatively little is known about the diet and feeding ecology of *Toxotes chatareus*, *Toxotes jaculatrix* or any toxotid.

For example, trophic level (express the relative position of an animal in the food webs) are critical to management are almost entirely lacking for *T. chatareus*, *T. jaculatrix* and up to now their position (trophic level) in the food web has never been studied in Malaysia, or elsewhere.

Therefore, the objective of this chapter is to compare the spitting success rate and accuracy of two species of archer fishes, and then determine if group size has an effect on spitting success rate and accuracy. Besides that, this chapter discusses the food composition of the two co-occurring and closely related archer species and determine whether they display different food preferences.

The Approach

Feeding Techniques

Four specimens – two *Toxotes chatareus* and two *Toxotes jaculatrix*, ranging in size from 7-8 cm were obtained from an aquarium store. They were kept in glass tanks (49 × 30 × 30 cm; length × depth × height) that were filled with brackish water to a height of 25 cm. Water temperature was kept at 26 °C, pH between 6.5-7.5 and NO_3 1-5 mg/L. Fluorescent lights were on from 8:00 AM to 10:00 PM. The tanks were aerated except during practice and test sessions, when pumps were turned off to ensure a calm water surface. Five combinations of the four fish were tested (1 = 1 *T. jaculatrix*; 2 = 1 *T. chatareus*; 3 = 2 *T. jaculatrix*; 4 = 2 *T. chatareus* and 5 = 2 *T. chatareus* and 2 *T. jaculatrix*).

For each group, the success per 100 spits was examined at five different heights (10, 15, 20, 25 and 30 cm above the water level), and live mealworms (2-2.5 cm) were used as prey items/targets. In the experiments the fish were fed only once daily, so that they were deprived of food for 24 h before practice and test sessions. When fish hit the target; small live mealworms were used as rewards. Less accurate fish were given enough food to bring their total daily ration up to the same as the most accurate fish. Thus, all fish were in the same state of hunger at the start of each day's testing session.

Targets (live mealworms) were offered on a loop of thin, nylon monofilament line, 5 cm in length. If the target was hit, it was knocked off the loop by the observer which then fell into the aquaria. The line was tied to a horizontal iron rod which was set into a vertical iron rod with height markings (cm), on the edge of the tank, at a height with intervals of 5 cm (Figure 4.1). The observer was seated in front of the aquarium and scored the spitting success rate and accuracy. A video camera placed at left angle in front of the aquarium was used to record each test (Figure 4.1). The cameras' light source was focused on the target so that its movement attracted the fish. The video camera was set to a fixed recording time of 30 minutes in order to avoid larger file sizes.

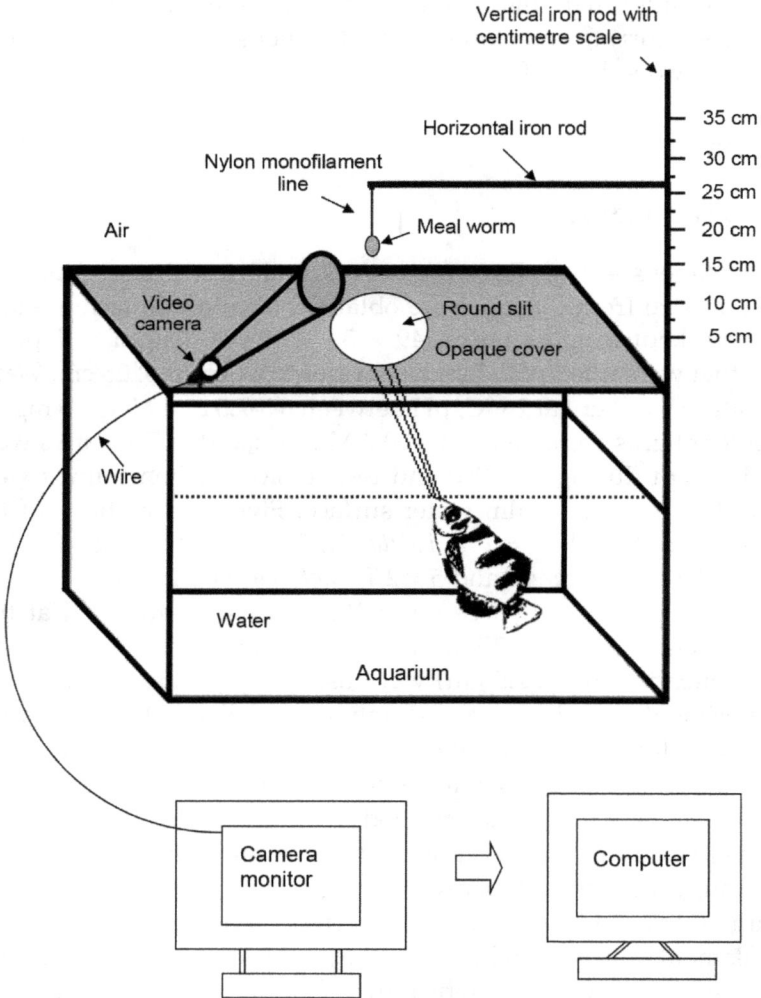

Figure 4.1. Schematic view of experimental setup for practice and tests of spitting success and accuracy of archer fishes in group and individually (or alone).

On the practice session, targets were offered at a height of 10 cm. If a fish squirted down the target within 5 minutes, during the next day's session the targets were offered 5 cm higher. If the fish did not succeed within 5 minutes, the targets were offered 5 cm lower, but not lower than 10 cm. Most of the fishes were able to hit the target at pre-selected heights within 5 minutes during the initial three months' practice session. Consequent to this, the test session commenced. During the test session similar target heights were used and targets that were hit and fell into the water were immediately eaten; targets that had not been hit were dropped into the water by the observer at the end of the 5 minutes' trial. During testing spitting success was scored as a direct hit to the target and accuracy which did not hit the target were categorised as too high, too low and to the right or left side of the target.

Analysis of variance (ANOVA) was performed on the spitting success data and Tukey's test was carried out to compare the means of test groups (1-5) and heights (10-30 cm) at $P < 0.05$.

On the other hand, a multinomial logit model was used to analyse the spitting accuracy data. The spitting accuracy were categorised as spitting right side high (RH), right side low (RL), left side high (LH), and left side low (LL). Since accuracies were polychotomous, no unique category can serve as a reference category. Therefore, any of the categories can serve as a baseline or comparison category. For each combination of values of group, height of target, and accuracy, the dependent variable was the log ratio of the probability of a spitting accuracy category compared to the probability of the reference category i.e., spitting left and low (LL). The log ratio was referred to as odds ratio. All statistical analyses were performed using MINITAB version 14 and SPSS version 15 software.

Stomach Content

A total of 128 fish (*T. chatareus*, n = 63, and *T. jaculatrix* n = 65) were collected using cast and three-layered trammel nets. After hauling, the catch was removed, and the fish were preserved in 10% (w/v) buffered formaldehyde for subsequent analysis in the laboratory. The preservative (4% (w/v) buffered formaldehyde) was also injected directly into the fishes' belly to prevent any further digestion and decomposition of the content.

In the laboratory, the digestive tract was removed and fixed in 70% ethanol to provide further and longer preservation in the museum.

Diet estimation was achieved by analyses of stomach contents from pre-selected samples (63 *T. chatareus* and 65 *T. jaculatrix*). Stomach fullness was determined according to Joyce et al. (2002). Stomach contents were analysed under the microscope and quantified in accordance with occurrence method (Gunn and Milward 1985; Hyslop 1980). Frequency of occurrence (*fo*) and percentage weight (wt %) were examined for different length classes.

Finally, diet composition data were also used for the estimation of the trophic levels of *T. chatareus* and *T. jaculatrix*. Trophic level (TROPH) expresses the position of organisms within the food webs that largely define aquatic ecosystems (Pauly et al. 1995, 1998; Pauly and Christensen 1995, 2000; Pauly and Palomares 2000). TROPH value was calculated from the dataset using TrophLab (Pauly et al. 2000), which is a stand-alone application for estimating TROPH and its standard error (S.E.) using the weight or volume contribution and the trophic level of each prey species to the diet (Pauly et al. 2001). Real consumers do not usually have TROPHs with integer values and the definition of TROPH for any consumer species *i* is:

$$TROPH_i = 1 + \sum_{j=1}^{G} DC_{ij} \times TROPH_j,$$

where $TROPH_j$ is the fractional trophic level of prey *j*, DC_{ij} represents the fraction of *j* in the diet of *i*, and *G* is the total number of prey species. Thus defined, the trophic level of aquatic consumers is a measurable entity that can take any value between 2.0 for herbivorous/detritivorous and 5.0 for piscivorous/carnivorous organisms (Pauly et al. 1998; Pauly and Palomares 2000).

The Outcomes

Feeding Techniques

In general, spitting by individuals within groups (groups 3-5) of two or more appeared more rushed and less controlled than spitting by fish kept alone (groups 1 and 2). Indeed, fish in groups had less time to take the best position as they were constantly fighting to take turns spitting and hence spitting was not as accurate. A detailed description of spitting success (hitting the target) and accuracy (not hitting the target) are presented in Table 4.1.

Table 4.1. Spitting success and spitting accuracy of archer fishes in different heights

Test groups	Replicates	Height (cm)	Spitting % success	Spitting % accuracy			
				RH	RL	LH	LL
1	a	10	40	21	21	16	2
	b	10	44	18	16	18	4
	a	15	32	24	20	19	5
	b	15	35	21	22	16	6
	a	20	27	27	25	11	10
	b	20	29	26	20	17	8
	a	25	28	24	22	15	11
	b	25	27	26	20	17	10
	a	30	20	27	26	17	10
	b	30	21	23	24	19	13
2	a	10	37	21	22	17	3
	b	10	42	20	18	15	5
	a	15	30	24	24	15	7
	b	15	31	23	20	16	10
	a	20	25	22	19	15	19
	b	20	27	25	20	12	16
	a	25	23	27	24	10	16
	b	25	28	25	21	12	14
	a	30	19	23	25	19	14
	b	30	18	24	26	16	16
3	a	10	29	23	22	19	7
	b	10	30	27	15	20	8
	a	15	26	24	28	15	7
	b	15	27	25	23	17	8
	a	20	20	24	26	16	14
	b	20	21	25	25	17	12
	a	25	17	27	25	17	14
	b	25	19	28	27	15	11
	a	30	14	26	25	18	17
	b	30	15	25	23	19	18

Table 4.1. (*Contd.*)

Test groups	Replicates	Height (cm)	Spitting % success	Spitting % accuracy			
				RH	RL	LH	LL
4	a	10	29	25	23	19	4
	b	10	27	26	17	22	8
	a	15	24	25	24	18	9
	b	15	25	25	20	22	8
	a	20	18	29	25	16	12
	b	20	20	32	23	17	8
	a	25	17	25	24	20	14
	b	25	18	30	20	17	15
	a	30	11	31	25	17	16
	b	30	13	25	23	19	20
5	a	10	25	26	23	20	6
	b	10	22	27	24	19	8
	a	15	17	26	27	18	12
	b	15	15	29	24	21	11
	a	20	12	25	28	19	16
	b	20	15	26	29	15	15
	a	25	11	32	28	14	15
	b	25	12	27	26	17	18
	a	30	10	27	29	19	15
	b	30	13	31	23	15	18

Test group 1: 1 *Toxotes jaculatrix*, 2: 1 *Toxotes chatareus*, 3: 2 *Toxotes jaculatrix*, 4: 2 *Toxotes chatareus*, 5: 2 *Toxotes chatareus* and 2 *Toxotes jaculatrix*.

Table 4.2 shows that the effect of test groups (groups 1 to 5) and heights (heights 10 to 30 cm) are significant ($P < 0.05$) in success of spitting. Although the interaction between test groups and heights are not significant (Table 4.2), the interaction plot is presented to show the patterns of percent (%) hits for the groups at different heights (Figure 4.2). Groups 1 and 2 consistently performed better than group 3 and 4 which in turn performed better than group 5 at all height levels.

Table 4.2. Analysis of variance on spitting success (%) for groups and heights

Source	DF	SS	MS	F
Groups	4	1483.6	370.9	131.52*
Heights	4	1697.6	424.4	150.5*
Groups×Heights	16	104.8	6.55	2.32
Error	25	70.5	2.82	
Total	49	3356.5		
	$r^2 = 97.9\%$	$r^2\,(\text{adj}) = 95.88\%$		

* Indicate significant difference at $P < 0.01$

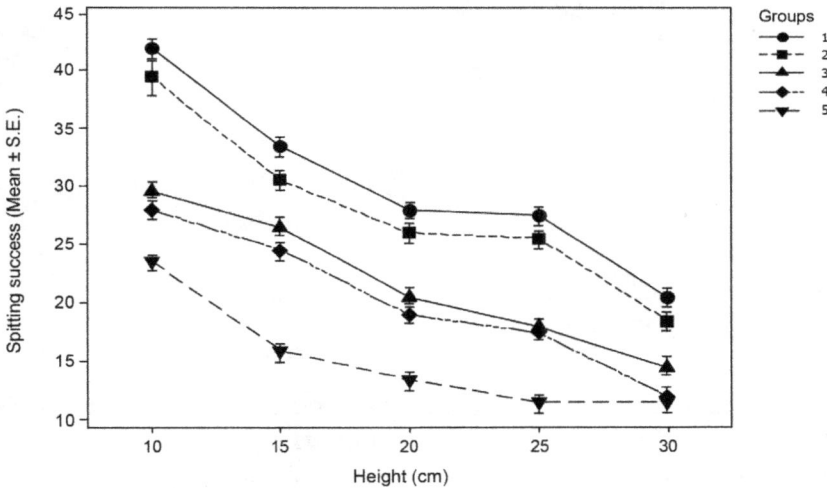

Figure 4.2. Plot of % hits (spitting success) for groups at different heights. Groups 1: 1 *Toxotes jaculatrix*, 2: 1 *Toxotes jaculatrix*, 3: 2 *Toxotes jaculatrix*, 4: 2 *Toxotes chatareus*, 5: 2 *Toxotes chatareus* and 2 *Toxotes jaculatrix*.

Multiple comparisons of means showed that groups 1 and 2 exhibited significantly high spitting success (group 1, LSM: 29.87%, and group 2, LSM: 27.55%) followed by group 3 (LSM: 21.45%) and group 4 (LSM: 19.79%). Test group 5 had the lowest spitting success (LSM: 14.88%) (Table 4.3). With regards to target height, the success of spitting was highest at 10 cm (LSM: 32.10%) and significantly higher than other target heights. Spitting success for target heights of 20 and 25 cm were not significantly different while spitting success at target height of 30 cm was significantly the lowest (Table 4.3).

Table 4.3. Least square means (LSM %) for groups and heights
on spitting success

Effects	LSM %
Groups	
1 (1 *T. jaculatrix*)	29.87[a]
2 (1 *T. chatareus*)	27.55[a]
3 (2 *T. jaculatrix*)	21.45[b]
4 (2 *T. chatareus*)	19.79[b]
5 (2 *T. jaculatrix* and 2 *T. chatareus*)	14.88[c]
Height (cm)	
10	32.1036[a]
15	25.8166[b]
20	21.0497[c]
25	19.5364[c]
30	15.1788[d]

[a,b,c,d] Different letters indicate significant
difference at $P < 0.05$

The analysis of spitting accuracy data indicated that the goodness-of-fit statistics for fitting the polychotomous logit model without the three-way interaction of group, height, and accuracy fit the model well. Both the G^2 and X^2 values were about 27.0 while the observed significance level was 0.99.

Based on the model, spitting accuracy categories RH, RL and LH showed no significant differences among the test groups. For instance, test group 1 was 1.23 times more likely to spit RH than LL compared to group 5 (Table 4.4). However, parameter estimates (λ) for group 1 as well as for other groups (2 to 5) were not significantly different from zero (Table 4.4).

The effects of target height on spitting accuracy categories indicated that the odds of shooting RH, RL and LH was about 1.5-2.7 times higher than LL for heights 10 cm and 15 cm compared to heights of 30 cm. However, at heights of 20 cm and 25 cm, the spitting accuracy categories were not significantly different compared to at 30 cm (Table 4.4).

Table 4.4. Parameter estimate, λ and estimated odds (e^{λ}) of the logit model on spitting accuracy

	RH	RL	LH	LL
		Accuracy categories		
Test groups				
1	0.209 (1.23)	0.170 (1.19)	0.296 (1.34)	0 (1)
2	-0.195 (0.82)	-0.305 (0.74)	-0.210 (0.81)	0 (1)
3	-0.091 (0.91)	-0.097 (0.91)	-0.028 (0.97)	0 (1)
4	0.000 (1)	-0.144 (0.87)	0.068 (1.07)	0 (1)
5	0 (1)	0 (1)	0 (1)	0 (1)
Height (cm)				
10	**0.836** (2.31)	**0.733** (2.08)	**0.989** (2.69)	0 (1)
15	**0.504** (1.66)	**0.403** (1.50)	**0.562** (1.75)	0 (1)
20	0.119 (1.13)	0.084 (1.09)	-0.015 (0.99)	0 (1)
25	0.173 (1.18)	0.088 (1.09)	-0.006 (0.99)	0 (1)
30	0 (1)	0 (1)	0 (1)	0 (1)

Bold letters indicate the parameter estimates are significantly different from 0 (zero)

With regards to success of spitting, test groups 1 and 2 scored higher than test groups 3 to 5. The highest spitting success recorded in the present study (*ca* 30%) was found to be lower in compassion with previous study (56%) (Timmermans and Vossen 2000) which is probably attributed to experiment set up. Lüling (1958, 1963) suggested that by spitting nearly vertically, the archer fish would, for the greater part, solve the refraction problem. However, Timmermans (1975, 2000), Timmermans and Vossen (2000) documented that archer fishes used a wide range of elevation angles (58-102°). They also reported that shooting success was not significantly different at elevation angles other than 90°. In the present study, relatively greater spitting success rate in *T. jaculatrix* at both group and individual level might be the reason as described by Lüling (1958, 1963) (i.e., *T. jaculatrix* can solve refraction problems squirting nearly vertically).

The present study shows that target height also affected the shooting success. At the height of 10 cm, the shooting success was higher in all groups. Accuracy that was observed at the height of 10 cm, could possibly be explained by the fishes' intention to jump evoked by the low target, and by the fact that jumps are about vertical. Archer fishes have been observed to jump at targets as low as about one body length (personal observation; Lüling 1958; Verwey 1928). In the present study, jumps were seldom observed and were not considered in the analysis.

Lüling (1958) suggested that the correct angle of elevation is easier to achieve than the correct angle in the horizontal plane (i.e., right left). However, in the present study no differences were observed in terms of direction (right or left) and distance (low or high) for spitting whereas for heights at 10 and 15 cm, the fish were more inclined to spit RH, RL, and LH compared to other heights.

Stomach Content Analysis

The stomachs of 63 *T. chatareus* samples were examined, and 8% was found empty and 92% contained food items, whereas in 65 *T. jaculatrix* stomachs 11% was empty and 89% contained food items. A total of 2 crustaceans (crab and shrimp), 4 different insect families, and 1 teleost species were identified (Table 4.5 and Figure 4.3). After grouping all food items into three categories, crustaceans in particular with the red clawed crab (*Sesarma bidens*) (70%, 67%) were the main prey items observed in both *T. chatareus* and *T. jaculatrix*, followed by insects (namely Formicidae 12%, 13%; Dytiscidae 5%, 6%; Araneidae 3%, 5%; and Cerambycidae 3%, 3%). *Penaeus* sp. occurred in 3% and 2% in *T. chatareus* and *T. jaculatrix* stomachs, respectively. Teleost (*Toxotes* sp.) constituted a tiny proportion of diet (5%) in *T. jaculatrix* while it was not found in *T. chatareus* stomachs (Figure 4.4). Crustaceans formed the majority of the diet, 73% (fo = 92) and 69% (fo = 89) by weight, followed by insects 23% (fo = 87) and 26% (fo = 69) in *T. chatareus* and *T. jaculatrix*, respectively (Table 4.5).

The estimated trophic level (TROPH) ranged from 3.230 to 3.590 with mean values of (3.422 ± 0.009) for *T. chatareus* and 3.240 to 4.390 and mean values of (3.420 ± 0.020) for *T. jaculatrix*. Trophic levels in both species gradually increased with size (Figure 4.5). Stomach content weight was observed to increase slightly with stomach size (Figures 4.6). Average stomach content weight for all stomachs (n = 128, *T. chatareus*, 63 and *T. jaculatrix*, 65) was 0.95 and 0.60 g, 1.5 and 1.6% of average body weight (65.09 and 67.97 g), respectively.

Figure 4.3. Major food items ingested by two archer fishes *T. chatareus* and *T. jaculatrix* collected from Sungai Santi, Johor coastal waters, Malaysia (a) Crab. Plus sign indicates chelate leg and asterisk indicates non-chelate leg; (b) shrimp; (c) insect; (d) teleosts.

Almost all archer fish stomachs were filled; only a small percentage of empty stomachs were encountered, which may reflect short periods of feeding as well as prolonged digestion. However, lengthy digestion itself has some amenities for dietary analysis: a larger proportion of certain prey species are identifiable. The two species of archer fishes in the Johor coastal waters feed primarily on crustaceans (mostly *Sesarma bidens*) and insects (in decreasing order of abundance: Formicidae, Dytiscidae, Araneidae and Cerambycidae) followed by teleost.

Table 4.5. Prey items observed in 128 archer fish (63 *T. chatareus* and 65 *T. jaculatrix*) stomachs from Sungai Santi, Johor coastal waters, Malaysia, grouped by major prey categories

Species	Prey category	N	W (g)	wt %	n	fo	w (g)
T. chatareus	Crustaceans	175	46	73	58	92	0.73
	Crab (*S. bidens*)	151	44	70	48	76	0.69
	Shrimp (*Penaeus* sp.)	24	2	3	10	16	0.03
	Insects	328	14	23	55	87	0.22
	Formicidae	153	7	12	29	46	0.11
	Dytiscidae	73	3	5	16	25	0.05
	Araneidae	57	2	3	6	10	0.03
	Cerambycidae	45	2	3	4	6	0.03
	Teleosts	0	0	0	0	0	0.00
	Total	503	60	96			0.95
T. jaculatrix	Crustaceans	184	27	69	58	89	0.42
	Crab (*S. bidens*)	156	26	67	50	77	0.40
	Shrimp (*Penaeus* sp.)	28	1	2	8	12	0.02
	Insects	205	10	26	45	69	0.15
	Formicidae	97	5	13	25	38	0.08
	Dytiscidae	51	2	6	10	15	0.03
	Araneidae	43	2	5	6	9	0.03
	Cerambycidae	14	1	3	3	5	0.02
	Teleosts	2	2	5	2	3	0.03
	Toxotes sp.	2	2	5	1	2	0.03
	Total	391	39	100			0.60

Crustacean (i.e. crabs and shrimps) dominated the stomach content by weight, while insects dominated the diet by number in all length classes of *T. chatareus* and *T. jaculatrix* (Table 4.5 and Figure 4.3). It is assumed that insects were captured by the two species by shooting from the overhanging mangrove vegetation in the study area. This feeding mechanism has been well documented (Rossel et al. 2002; Schuster et al. 2004, 2006; Timmermans 2000; Timmermans 2001; Timmermans and Vossen 2000). Smaller individual with total length

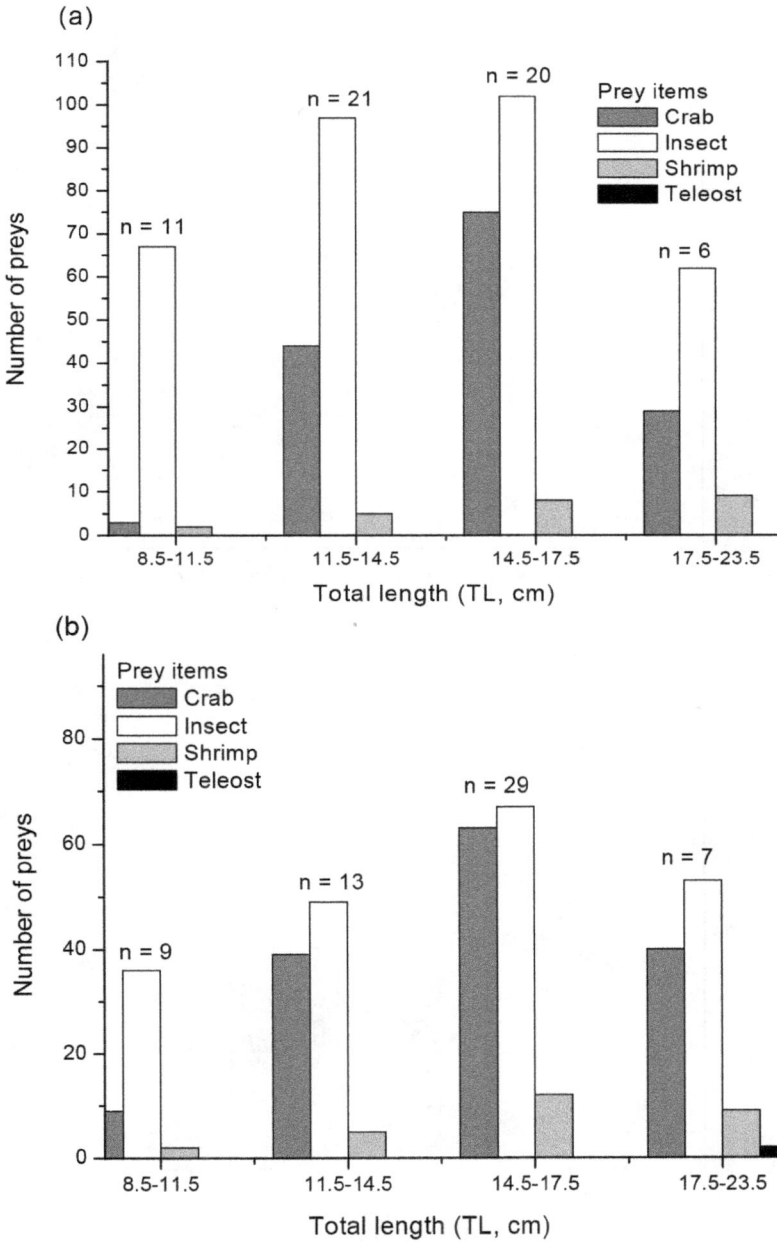

Figure 4.4. Number of food items ingested by (a) *Toxotes chatareus* and (b) *Toxotes jaculatrix*. *n*: number of fish in each length class (excluding empty stomach).

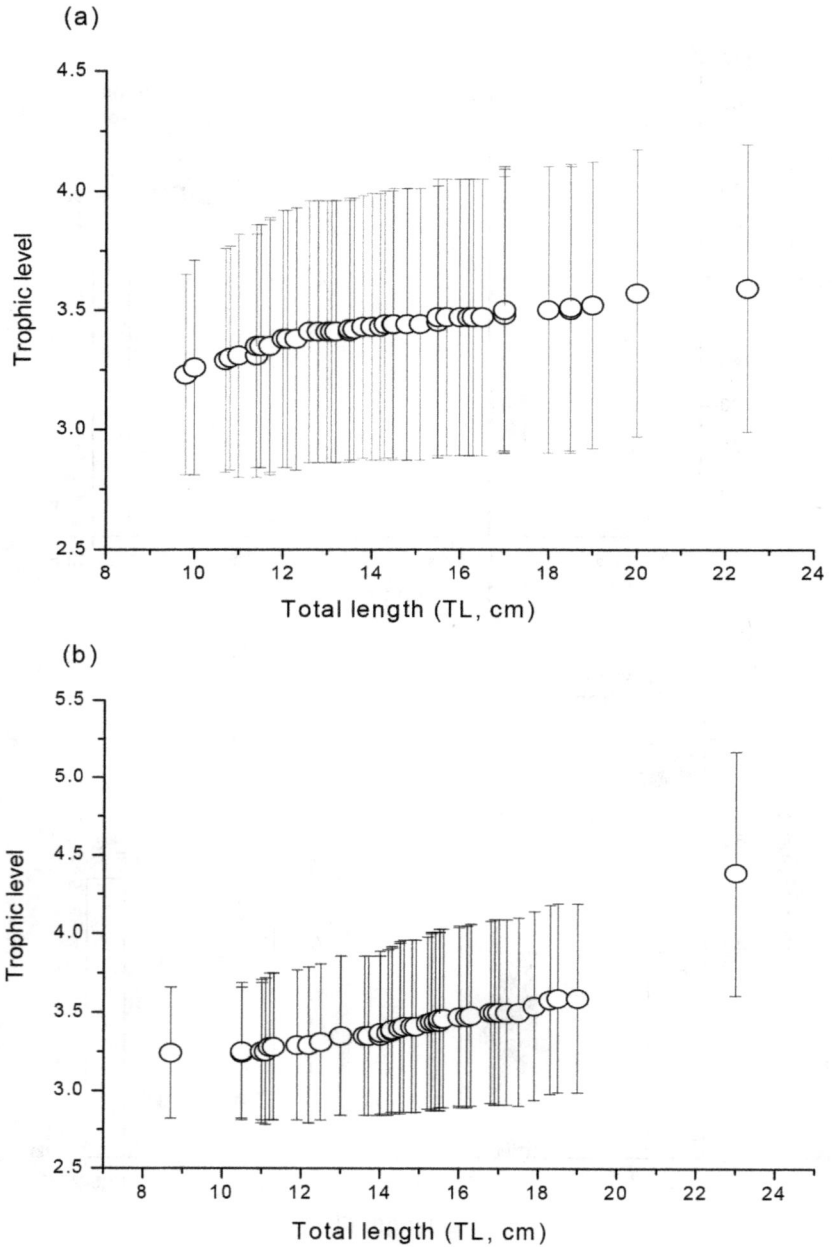

Figure 4.5. Identification of the trophic level of (a) *Toxotes chatareus* and (b) *Toxotes jaculatrix* (open circles represent mean TROPH values/trophic level and solid bars represent variants of prey items).

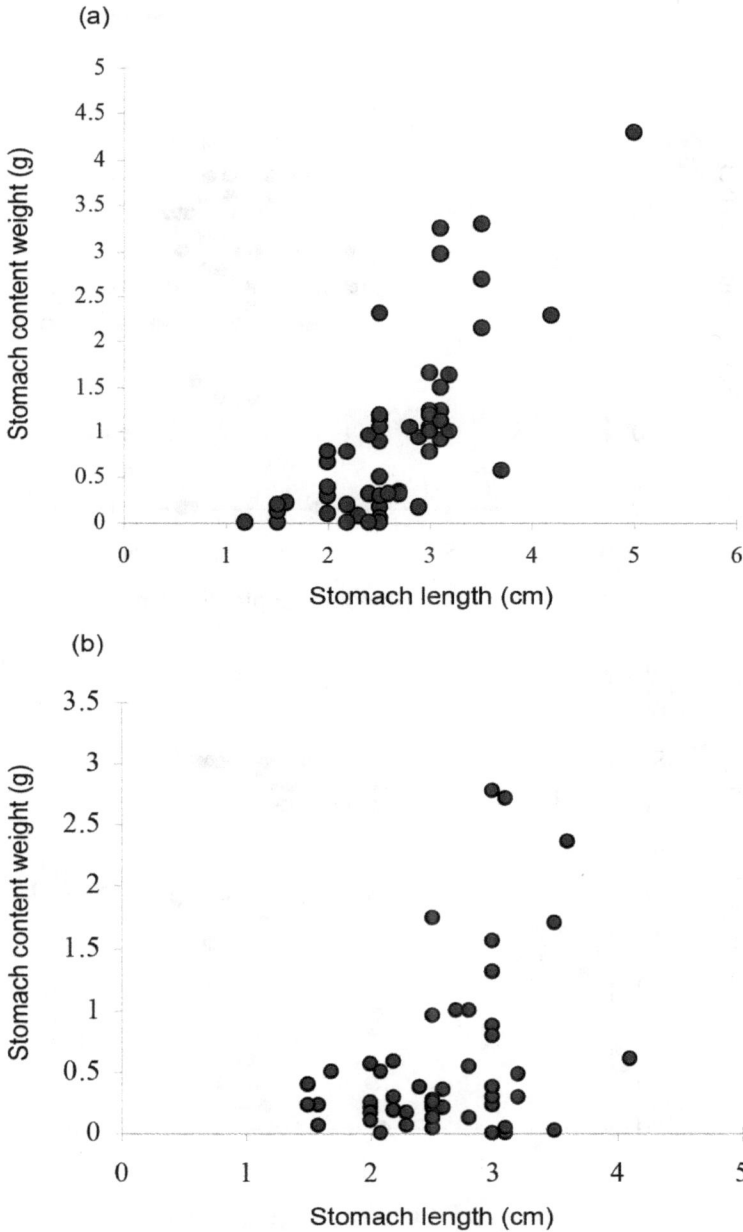

Figure 4.6. Stomach length and preserved weight of stomach contents. (a) *Toxotes chatareus* and (b) *Toxotes jaculatrix* (closed circle represent individual sample; sample size for both species, $n = 53$).

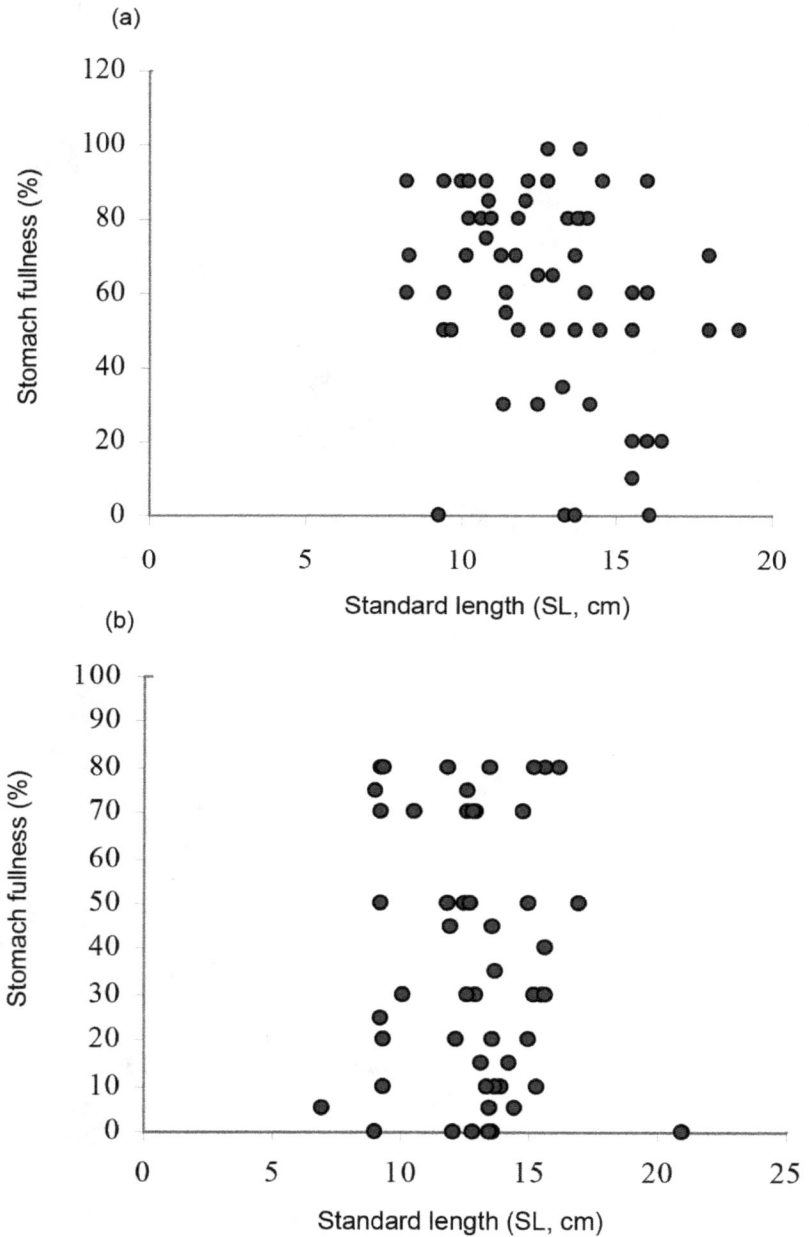

Figure 4.7. Stomach fullness in (a) *Toxotes chatareus* and (b) *Toxotes jaculatrix* in relation to standard length (closed circle represents individual sample; sample size for both species, $n = 58$.

of 8.5-11.5 cm of both species fed on a smaller number of prey items compared to larger length classes.

The diet differences among the size classes are probably due to the energy requirements, which vary according to the developmental stage. Indeed, during ontogeny, fish often change their diet mediated by allometric, morphological changes (Karpouzi and Stergiou 2003), thus being able to exploit sequentially a series of prey sizes ranging from phytoplankton and small size zooplankton to much larger prey (Wootton 1998). In the present study, the results show that *T. chatareus* and *T. jaculatrix's* diet includes crabs, shrimps, and insects prey, which is consistent with Blaber (2000). However, Blaber found that *T. chatareus* also take plant material, which he explained as an occasional behaviour. In addition, the similar ontogenetic changes in the diet in *T. chatareus* were reported in the present study and by Blaber. The methods of stomach fullness assessment in the present study allowed to observe the prey items in the stomach by dissection. A similar method of assessment has been also reported by Joyce et al. (2002). Stomach fullness was maximal in standard length (SL) of 10-15 cm fishes whereas lesser in SL of >15 cm fishes; it is suspected that a high rate of digestion may be involved in this observation (Figure 4.7). Larger archer fishes appear to become more carnivorous with growth, capable of capturing larger crabs, shrimps and insects while also exercising some cannibalism.

The estimated TROPH value for *T. chatareus* and *T. jaculatrix* are similar to that calculated by Stergiou and Karpouzi (2002) for other pelagic species such as *Gadiculus argenteus argenteus*, *Trachurus mediterraneus*, and *Trachurus trachurus*. All these species exhibit similar feeding preference, namely, crustaceans, fish larvae, and mysids, and can be considered as largely carnivores. The results from the present study indicate that the two species of archer fish are largely opportunistic predators (with a few exceptions) feeding on a wide spectrum of prey species such as crustaceans, insects, and, rarely, teleost, depending on the food availability in the environment.

Conclusion

In this chapter, the video footage enabled to measure spitting success and accuracy. There were no differences between the groups and individuals with respect to the position from where spitting took place (distance from the target's perpendicular), and with respect to the spitting angle (because the set up allowed them to spit only over a very narrow range of angles). The data also illustrates that the two species of archer fishes (*T. chatareus* and *T. jaculatrix*) did not follow any patterns in spitting to knock down their preys from outside water environment. These findings suggested that individual specimens kept alone of both species are more successful at hitting targets at any height compared to groups. In conclusion, this chapter has critically examined for the first time the spitting performance of two closely related archer fishes (*T. chatareus* and *T. jaculatrix*) in groups and individually at different target heights. These findings would be useful for the fisheries' biologists to understand the feeding behaviour of these fascinating fishes in the wild. Since no studies on stomach content analysis and trophic level of archer fishes have been conducted in Johor coastal waters or elsewhere, this data will be useful in ecological modelling for a better representation of the trophic flows associated with large, medium, and small pelagic fish.

Gastric Emptying and Digestion of Archer Fish

Archer fish is carnivorous in nature and consumes wide range of preys, including terrestrial insects, shrimps and teleost. To achieve optimal digestion and feed utilization in fish, the food needs to be palatable and of appropriate size. The rate of energy supply from all nutrients depends on the animal's ability to digest the food ration at appropriate rates, and to extract the maximum possible levels of energy and nutrients from the food for its metabolic requirements. Nutrient digestibility of fishes has been measured in different ways. Inert marker(s) has mostly been used by the researchers to calculate apparent digestibility (absorption) coefficients (ADCs) by monitoring relative changes in inert marker and nutrient concentration in faeces relative to the diet fed.

To date there is no information on gastric emptying rate and nutrient absorption along the digestive tract of *Toxotes jaculatrix*. Therefore, the present chapter represents a modelling approach to indirectly describe the digestion process in archer fish *T. jaculatrix*. This can be achieved by tracing the movement of food item in the gut passage using X-radiography and gastrectomy techniques. The absorption of macronutrients passing through the alimentary tract of *T. jaculatrix* conditioned to feed on whole mealworm were also studied.

The Approach

A total of 160 fish were collected using cages, cast, three layered trammel nets and hand line (rod fishing). Samples were separated into two groups (i.e., live samples n = 50 and dead samples n = 110) and transported to the laboratory. In the laboratory, the pre-selected dead samples of various sizes were sampled by removing the alimentary tracts from the carcasses. The tracts were significantly divided into four sections, as depicted in Figure 5.1, and their contents were subsequently extruded. Due to the small size of the digestive tract and several analyses required, the contents from each part of the digestive tract i.e., stomach (a), anterior intestine (b), mid intestine (c), and rectum (d), were pooled into labelled small plastic bottles (urine bottle) and frozen at −16°C. In contrast, the live samples were divided into different group's base on their size classes in the 50 litre (L) holding tanks. The tanks were equipped with circular air stones. Sea water was recirculated through a biofilter column at 15 L min^{-1} with replacement of new seawater from the reserve tanks (300 L)

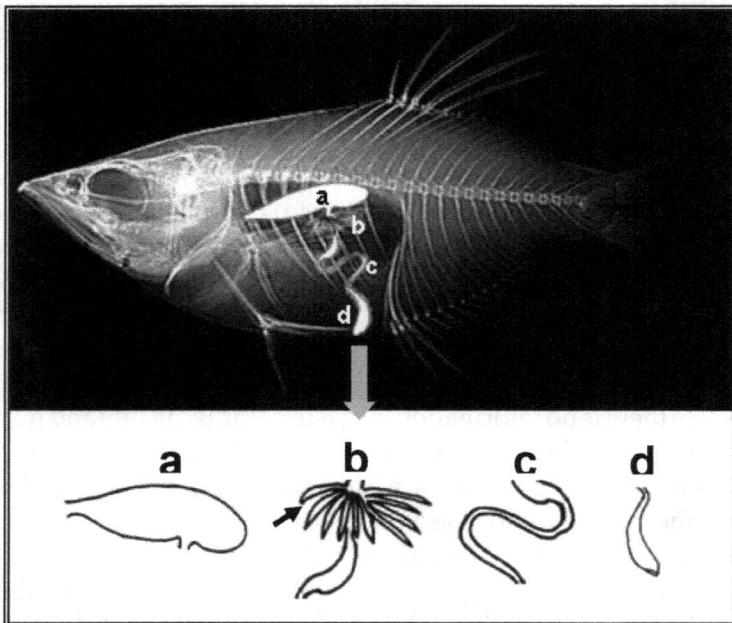

Figure 5.1. Diagram of the archer fish alimentary canal/tract. Four sectioned of alimentary tract: (a) stomach; (b) anterior intestine (black arrow indicate pyloric caeca); (c) mid intestine; (d) rectum (posterior intestine).

at 1.1 L min[-1]. The effluent water was pumped to the top of a cylindrical tank and passed through the biofilter (fiber) and polypropylene ring, which provided a substratum for the growth of microflora and macroflora and filter feeder fauna, essential for the functioning of the biological filter. Ammonia level was maintained at 1.5 mg/L. There was a temperature control unit; therefore, the water temperature was maintained and fixed at 27°C. In the holding tank, fish were fed mealworms (size 2.2 ± 0.001 cm). Live fishes were transferred to the experimental tank (n = 10 fish/tank) and accustomed to accept whole mealworms meal over a two-week period. Mealworms were frozen and freeze dried and injected with barium sulphate paste (Conc. 1 g $BaSO_4$: 5 ml distilled water) at approximately 0.1-0.15 ml/g wet weight and again kept frozen prior to the start of the X-ray experiment. It is important to use amounts of $BaSO_4$, in each mealworm, which is proportional to its weight to produce similar image densities as the food breaks up. Conversely, in the gastrectomy experiment similar size live fresh mealworms (without $BaSO_4$ paste) were used as food item.

All fish were deprived of food for 120 hr and then offered pre-weighed labelled mealworms (injected $BaSO_4$ mealworm for X-ray samples and fresh live mealworm for gastrectomy samples) to satiation (when offered food was no longer eaten by fish). The exact amount eaten was recorded for each fish. At a selected time after feeding (6, 12, 24, 36, 48, 60, 72 and 96 hr), the fish were anesthetised using 0.22 ml L[-1] of TRANSMORE® (NIKA) for 10-15 min and immediately X-rayed with a microradiographic unit (M60, Softex, Tokyo, Japan) to trace the movement of food along gut (Al-Aradi 1986; Mazlan and Grove 2003). The remnant 40 fish were killed by a blow to the head at similar time intervals and alimentary tract opened to remove contents of each fish and the wet weight of content was taken to the nearest 0.001 g. Faeces were collected twice daily by siphoning, sieved and frozen at −16° C and kept in plastic bottle for further chemical analysis.

A total of 40 *T. jaculatrix* samples (gastrectomy samples) were used for gastric emptying study. Curves of the change in contents with time were fitted to the data using the maximum voluntary meal size as a standard, as used by Fletcher et al. (1984) and Mazlan (2001). The square root model (Andersen 1998, 1999, 2001; Andersen and Beyer 2005; Mazlan 2001) used to describe gastric emptying of the test meal was as:

$$S_t = S_0(1 - S_0^{(\alpha-1)})\rho(1 - \alpha)t)^{(1-\alpha)^{-1}}) + \xi$$

where S_t is the total stomach content at time t after ingestion of a meal of size S_0 (stomach content at time zero) and ξ is the random error term. Model parameters for a maximum satiation meal (when S_0 = maximum satiation amount S_{max}) and the rate constant, ρ gastric emptying rate (GER), and gastric emptying coefficient α were estimated by non-linear iterative method using Levenberg-Marquardt algorithm.

Samples of diets and gastrointestinal contents were weighed to the nearest mg and moisture content was determined by measurement of weight loss after drying the samples in an ALPHA 1-2/LD PLUS freeze dryer at –40 to –60°C for 3-4 days depending on the size of the samples. The dried samples were ground to homogenous powder using pestle and mortar and kept in small glass vial in desiccators for further biochemical analyses. The inorganic matter or ash content of samples was determined for each part of gut samples using a Wisetherm® DH-WFMH0213 furnace at 450°C for 3.5 hr. The ash content was calculated by the difference between initial and final mass of the sample. C.H.N.S. Eager 300 analyser was used to determine total nitrogen content. Protein was calculated from nitrogen content multiplied with the factor of 6.25 (Mazlan and Grove 2003). The lipid (fat) of different gut content was determined by lipid analyser in a Soxtec 2055 Avanti Foss lipid analyser. Carbohydrate levels of dry weight were estimated by the formulation: % Carbohydrates = 100 – (% protein + % lipid + % ash) (Xin et al. 2008). Energy content (gross energy, GE) of the samples was determined by bomb calorimeter in an IKA® C-200 bomb calorimeter.

The nutrient contents in the digesta were used to determine the relative percentage absorption gradient (using ADCs) along the alimentary tract, based on ash contents of adjacent samples following the formula suggested earlier by Conover (1966) and later modified by Kolb and Luckey (1972), Maynard et al. (1969) and Mazlan and Grove (2003):

$$\text{ADCs (\%)} = 100 \times [1 - (\text{Ash}_a : \text{Nutrient}_a \div \text{Ash}_b : \text{Nutrienta}_b)].$$

To model the gastric emptying curve, a non-linear regression equation was fitted using Microcal Origin™ version 6 software and the parameters to be estimated were named and declared. The non-linear procedure first examines the starting value specifications of the parameters and evaluated the Chi square (X^2) value at each combination of values to determine the best set of values to start the iterative algorithm. The programme uses the Levenberg-Marquardt iterative method (Mazlan 2001).

The Outcomes

The movement and conditions of food items in alimentary tract after different time intervals since feeding are described in Figure 5.2. The X-radiographic images show that within 6-12 hr after feeding, the stomach is considering full with the residuum and a small portion of food has entered the proximal and distal arms of the intestine loop and partly has reached the rectum; 24, 36 and 48 hr since feeding, approximately 10-20% of the original meal still remains in the stomach (Figure 5.2). After 60 and 72 hr post feeding, less than 5% of the original

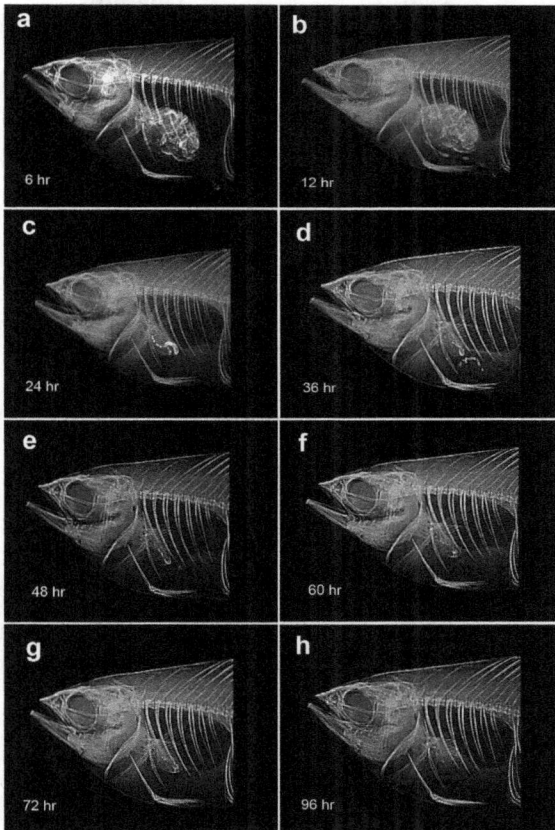

Figure 5.2. Gastric emptying in archer fish (*Toxotes jaculatrix*), voluntary fed to satiation (8-13% BW) with whole mealworms (containing 0.1-0.15 ml BaSO$_4$/g mealworms) as a disperse type radio opaque marker. (a-h): X-radiographic images traced food movement in digestive tract after different times since feeding.

meal remains in the stomach. Finally, within 96 hr after feeding, the food items have completely left the stomach and have concentrated in the posterior part of the intestine and rectum, ready to be defecated (Figure 5.2). Similar results were observed in gastrectomy method for the 40 fishes (Table 5.1). Voluntary satiation feeding in these fish ranged between 8 and 13% of their body weights.

Table 5.1. Amount fed (S_0) and residuum in stomach after gastrectomy at stated time since feeding for *Toxotes jaculatrix* fed with whole mealworms. Note that temperature was fixed 27°C in the experiment

Fish ID	TL (cm)	Fish wt. (g)	S_0 ($t = 0$) (g)	Time since feeding (hr)	Stomach residuum (g)
1	23.00	275.31	23.80	6	21.00
2	19.50	223.41	22.56	6	20.00
3	19.75	227.67	23.35	6	22.00
4	19.00	196.78	23.21	6	19.00
5	19.00	213.67	23.18	6	18.50
6	18.90	189.06	21.05	12	19.00
7	18.50	176.35	21.01	12	17.80
8	18.45	167.26	21.00	12	18.00
9	18.30	157.34	21.00	12	17.15
10	17.90	139.04	20.00	12	17.35
11	17.34	121.41	19.43	24	11.10
12	17.20	122.13	19.52	24	11.21
13	17.20	118.32	18.00	24	12.01
14	17.00	113.00	18.07	24	12.10
15	17.00	111.67	17.00	24	12.00
16	17.90	120.12	18.61	36	7.12
17	16.90	102.78	14.11	36	7.22
18	17.00	112.05	15.21	36	7.00
19	16.80	111.76	14.05	36	7.45
20	16.00	101.67	13.01	36	7.50
21	15.20	104.23	13.70	48	4.50
22	14.90	98.03	10.90	48	4.00
23	15.03	105.13	10.23	48	4.12
24	14.56	96.12	10.21	48	3.05

25	14.30	86.91	10.01	48	3.00
26	14.00	78.03	9.80	60	1.35
27	13.98	63.21	9.65	60	1.65
28	13.54	64.07	9.06	60	2.00
29	14.00	72.90	9.70	60	1.20
30	13.12	59.23	7.56	60	1.12
31	12.60	45.73	6.80	72	0.65
32	12.50	43.45	6.40	72	0.60
33	12.32	42.32	6.20	72	0.50
34	12.20	47.12	6.73	72	0.60
35	12.13	42.05	5.89	72	0.42
36	12.12	56.03	8.67	96	0.00
37	12.10	46.34	6.90	96	0.00
38	12.00	43.11	4.87	96	0.00
39	12.00	43.12	4.79	96	0.00
40	12.00	46.02	6.01	96	0.00

However, the fish did not all eat the same amount. To combine the results, the two points representing the largest voluntary maximum meal size of an archer fish *T. jaculatrix* at time zero (S_0) were taken as the best estimate of maximum satiation amount (S_{max} = increased allometrically with fish weight) (Figure 5.3) and its stomach residuum (S_t) at stated time (t) after feeding was plotted graphically by non-linear regression using square root model (Andersen 1998, 1999, 2001; Andersen and Beyer 2005; Mazlan 2001). The estimated rate parameters (gastric emptying rates) are shown in Figure 5.4. The average maximum meal size ($S_{0/max}$) for archer fish (*T. jaculatrix*) used in the present study was 24.1 g and the gastric emptying rate parameter (ρ) was 0.12 g hr^{-1} and gastric emptying coefficient (α) = 0.5. The amount of stomach residuum 6 and 12 hr after feeding was approximately 20 and 18 g and this value dropped steadily with time up to 24 hr (Figure 5.2). The emptying process observed in the present study was closely similar to that of square root model as reported for other carnivorous fishes (Andersen 1998, 1999, 2001; Andersen and Beyer 2005; Mazlan 2001), except for the first stages of GER. The prey mealworm has thick chitinous layer, which in turn influences the packing of the mealworm in the stomach and may prolong the digestion period.

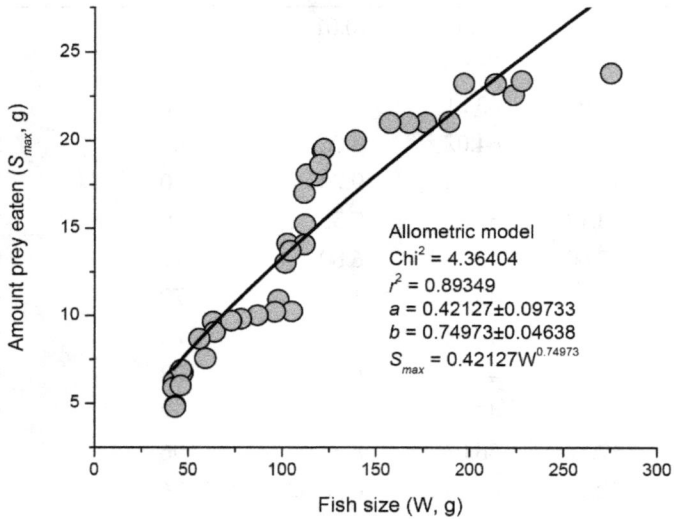

Figure 5.3. Relationship between amount of satiation meal (S_{max}, g wet weight) and fish size (wet weight, g) of archer fish *Toxotes jaculatrix* feeding on mealworm in fixed water temperature at 27°C.

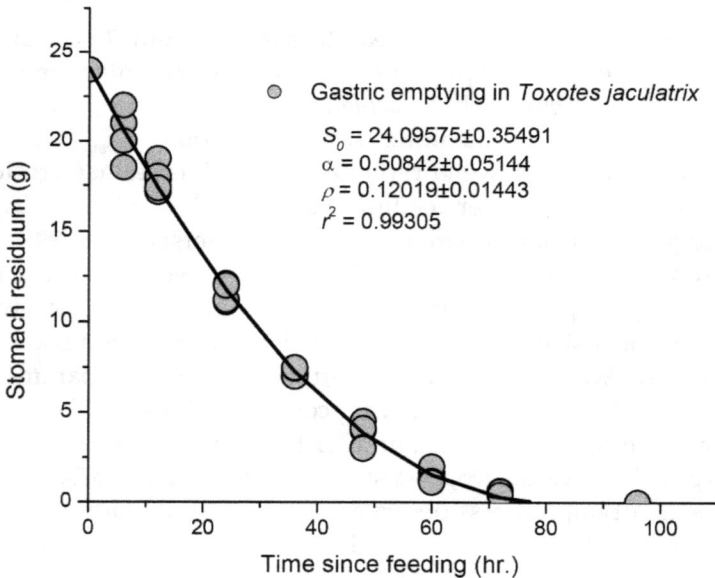

Figure 5.4. Gastric emptying curve of *Toxotes jaculatrix* feeding on whole mealworm fitted using square root model $[S_t = S_0(1 - S_0^{(\alpha-1)}\rho(1-\alpha)t)^{(1-\alpha)^{-1}}) + \xi]$ where S_0 = maximum meal size at time zero, ρ = gastric emptying rate (GER) and α = gastric emptying coefficient.

The mean ash content of experimental meal (mealworm), and digesta of wild and laboratory reared archer fish (*T. jaculatrix*) are depicted in Tables 5.2 and 5.3. Mealworms contained comparative lesser amounts of ash (15.8%) and moisture (66.8%) content than faeces (ash 76.9%, moisture 76.0%) ($P < 0.05$) (Table 5.2). Stomach of both wild (39.3%) and laboratory (37.9%) archer fish contained comparatively less ash content than anterior (43.6%, 49.2%) and posterior intestine (57.9%, 74.3%, respectively) (Table 5.3). In the wild archer fish, the ash content had risen from 39.3-43.6% of the sample dry mass (DM) by the time food reached the anterior intestine and increased this level thereafter (Table 5.3a). Similarly, ash concentration in laboratory fish rose progressively towards posterior regions of the gut to reach around 74.3% (Table 5.3b). The moisture content in laboratory-reared archer fish decreased over time along the gut, ranging from 72.0% to 45.0% (Table 5.3b). Similar trend was observed in wild samples where moisture content decreased from 49.9-35.7% along the gut (Table 5.3a). However, the concentrations of ash and moisture at various times after feeding with mealworm in the laboratory confirm that these were remarkably constant for a stated region (Figure 5.5b).

Protein was by far the largest fraction of the organic matter present in the experimental meal mealworms dried sample (57.2%), which contained comparatively lesser amount of lipid (26.0%) and very small amount of carbohydrate (0.86%). Whereas in faeces the concentration of protein, lipid, and carbohydrate was 18.9%, 3.1% and 0.93% respectively (Table 5.2). Stomach contained comparatively higher protein (49.9%, 48.7%) than other parts of the digestive tract (e.g., anterior intestine 45.7%, 41.0%; mid intestine 39.8%, 33.5%) in both wild and laboratory reared samples respectively (Table 5.3). Similarly, stomach contained relatively higher lipid content (9.8%, 12.6%) than anterior (9.3%, 8.9%) and mid intestine (7.9%, 6.9%) for both wild and laboratory samples in that order (Table 5.3). In laboratory fish, carbohydrate fractions were normally less than 1% of DM for all portions of the digestive tracts, whilst it was relatively high in anterior (1.2%) and posterior (1%) intestine of wild samples (Table 5.3).

The concentrations of each nutrient (except the carbohydrate in anterior gut section, $P < 0.05$) within a stated gut section showed that there were no significant differences over time ($P > 0.05$) (Figure 5.5). However, nutrient contents were found to decrease progressively along the gut at all times (Table 5.3b). These declines were significant for protein, for total lipids, digestible energy, and carbohydrate

Table 5.2. Proximate analyses of experimental meals (% DW ± S.E.)

Samples	Ash	Moisture	Protein	Total lipid	Carbohydrate	Energy (kJ g^{-1})
Mealworm (38.54± 0.15 g, n=600)	15.84[a]±0.57 (n=6)	66.88[a]±0.58 (n=6)	57.29[a]±0.51 (n=6)	26.01[a]±0.23 (n=6)	0.86[a]±0.01 (n=6)	26.72[a]±0.23 (n=6)
Faeces	76.93[b]±0.31 (n=6)	76[b]±0.52 (n=6)	18.98[b]±0.34 (n=6)	3.17[b]±0.13 (n=6)	0.93[a]±0.04 (n=6)	4.89[b]±0.56 (n=6)

Mean values within the same column having the same superscript are not significantly different (*P* > 0.05); n: number of replicates analysed

Table 5.3. Proximate analysis of digesta (% DW ± S.E.) from stated alimentary tract sections of archer fish (*T. jaculatrix*) (a) samples from the wild and (b) laboratory-held archer fish fed on whole meal

(a) Proximate analysis of digesta (wild archer fish)

	In. com. (Ash)	Org. com.			
		Protein	Total lipid	Total CHO	Total eng. (kJ g^{-1})
Test meal (mealworm)					
Stomach	39.39± 0.25 (n=3)	49.92± 0.25 (n=3)	9.84± 0.52 (n=3)	0.85± 0.01 (n=3)	23.77± 0.35 (n=3)
Pyloric caeca (Ant. int.)	43.69± 0.57 (n=3)	45.79± 0.25 (n=3)	9.32± 0.25 (n=3)	1.20± 0.02 (n=3)	22.76± 0.25 (n=3)
Intestine (Mid int.)	51.34± 0.36 (n=2)	39.81± 0.44 (n=2)	7.95± 0.26 (n=2)	0.90± 0.01 (n=2)	20.98± 0.36 (n=2)
Rectum	57.91±0.26 (n=2)	35.79± 0.25 (n=2)	5.3± 0.36 (n=2)	1.00± 0.01 (n=2)	21.2± 0.51 (n=2)

(b) Proximate analysis of digesta (laboratory reared archer fish)

	In. com.		Org. com.			
	Ash	Moisture	Protein	Total lipid	Total CHO	Total eng. (kJ g^{-1})
Test meal (mealworm)	38.93±0.12 (n=6)	75.29[a]± 0.23 (n=6)	49.35[a]± 0.86 (n=6)	11.12[a]± 0.56 (n=6)	0.60[a]± 0.01 (n=6)	13.23[b]± 0.47 (n=6)
Stomach	37.92[a]± 0.51 (n=2)	72.03[b]± 0.36 (n=2)	48.78[a]± 0.25 (n=2)	12.63[a]± 0.25 (n=2)	0.67[b]± 0.02 (n=3)	13.28[b]± 0.24 (n=3)
Pyloric caeca (Ant. int.)	49.21[b]± 1.48 (n=2)	59.7[c]± 0.30 (n=2)	41.04[b]± 0.16 (n=2)	8.97[b]± 0.25 (n=2)	0.78[c]± 0.02 (n=2)	15.73[c]± 0.25 (n=2)
Intestine (Mid int.)	58.73[c]± 0.25 (n=2)	52.97[d]± 0.25 (n=2)	33.54[c]± 0.25 (n=2)	6.98[c]± 0.25 (n=2)	0.74[d]± 0.02 (n=2)	11.28[a]± 0.25 (n=2)
Rectum	74.36[d]± 0.51 (n=2)	45.02[e]± 0.51 (n=2)	23.86[d]± 0.25 (n=2)	1.12[d]± 0.01 (n=2)	0.66[ab]± 0.02 (n=2)	3.24[d]± 0.25 (n=2)

Mean values within the same column having the same superscript are not significantly different ($P > 0.05$) (In. com.: Inorganic compound; Org. com.: Organic compound; Total eng.: Total energy; Ant. int.: Anterior intestine; Mid Int.: Mid intestine)

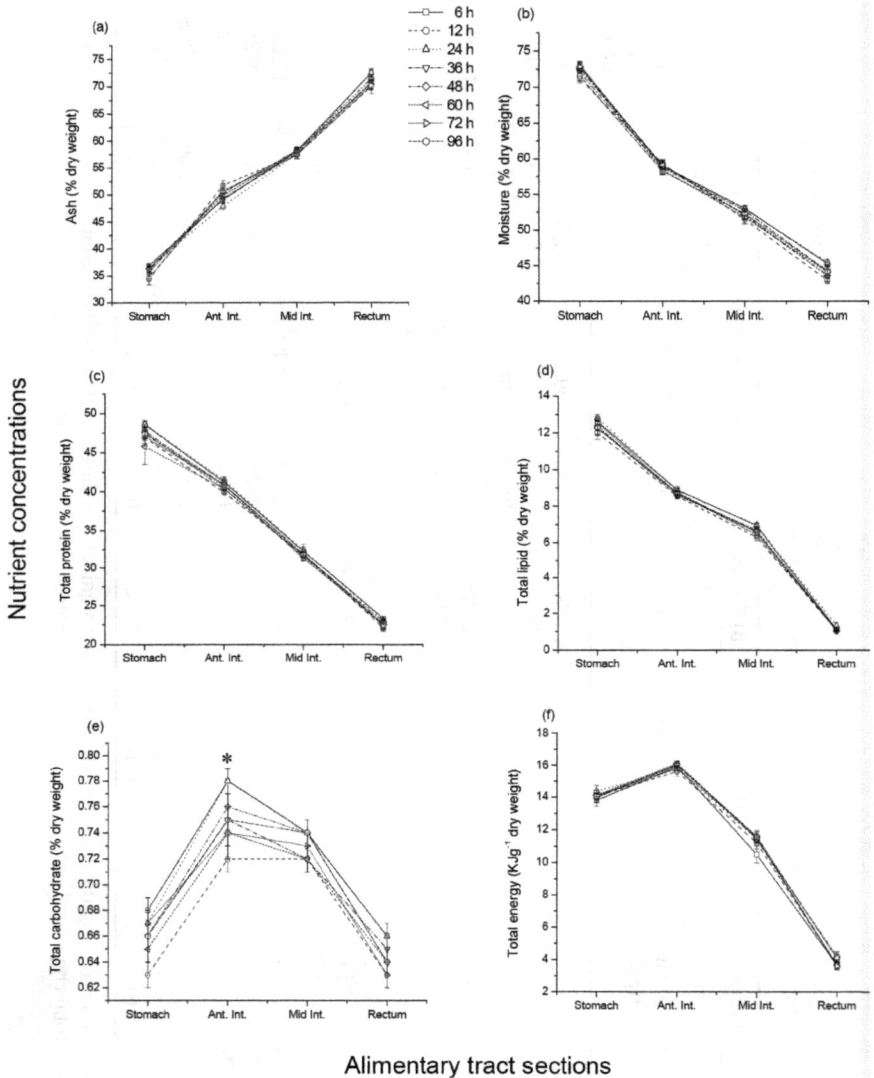

Figure 5.5. Macro nutrients (a: ash, b: moisture, c: total protein, d: total lipid, e: total carbohydrate; and f: total energy) concentrations in digesta from different gut sections at different times after feeding whole mealworms in laboratory-held archer fish. Vertical bar denoted standard error and asterisks indicate significance level at $P < 0.05$ (Ant. int.: Anterior intestine; Mid int.: Mid intestine).

($P < 0.05$) (Table 5.3b). Tamhane's procedure for multiple comparisons was used to locate where changes in content occurred. For protein, contents of adjacent sections of the gut were not significantly different, but the drop in protein was real by the time food reached the mid intestine and rectum; the same was true in the decline between the anterior intestine and rectum. Similar results were also observed for total lipids and digestible energy contents (Table 5.3b). Carbohydrate contents changed in a different manner along the gut. A significant increase occurred between stomach and anterior intestine, followed by a clear decrease in the mid intestine and rectum.

The laboratory study was limited to the non-natural diet of mealworm, because of the feeding behaviour as well as limited numbers of archer fish and time constraints. The stomach contents (section a) showed no consistent change over 96 hr, suggesting that there was little selective separation of nutrients (protein and lipid) from food which had not yet cleared the stomach.

The mean energy content for whole mealworm and faeces was 26.7 ± 0.2 kJ g^{-1} and 4.8 ± 0.5 kJ g^{-1} respectively ($P < 0.05$) (Table 5.2). There was a significant difference in energy contents of different portions of the gut contents in laboratory samples ($P < 0.05$) (Table 5.3b). For example, the anterior intestine contents had significantly higher energy density (15.7 ± 0.2 kJ g^{-1}) than their mid (11.2 ± 0.2 kJ g^{-1}) and posterior intestine (3.2 ± 0.2 kJ g^{-1}) contents ($P < 0.05$) (Table 5.3b). Similar trend also observed in the wild samples (Table 5.3a).

Figures 5.6a-d show estimates of apparent digestibility coefficients (% ADCs) for macronutrients and energy as food passes through the different gut sections independent of time since feeding. For wild archer fish, whatever the original diet mix, approximately 17.3% of protein, and 15.0% of total lipids and 27.3% of carbohydrate appear to have been absorbed in passage between the stomach and the anterior intestine (including the pyloric caeca). There was an unexpected negative apparent 'absorption' between samples from the stomach- and anterior-intestine for carbohydrate analysed (Figure 5.6c). In the passage of the remaining nutrients from mid-intestine to rectum, approximately 20.4% of protein, 41.0% of total lipids and 2.0% of carbohydrate were absorbed.

The laboratory held samples fed on non-natural diet whole mealworm exhibited similar patterns of apparent absorption efficiencies for the major nutrients when compared with wild samples (Figures 5.6a-d). The difference from wild samples was that active

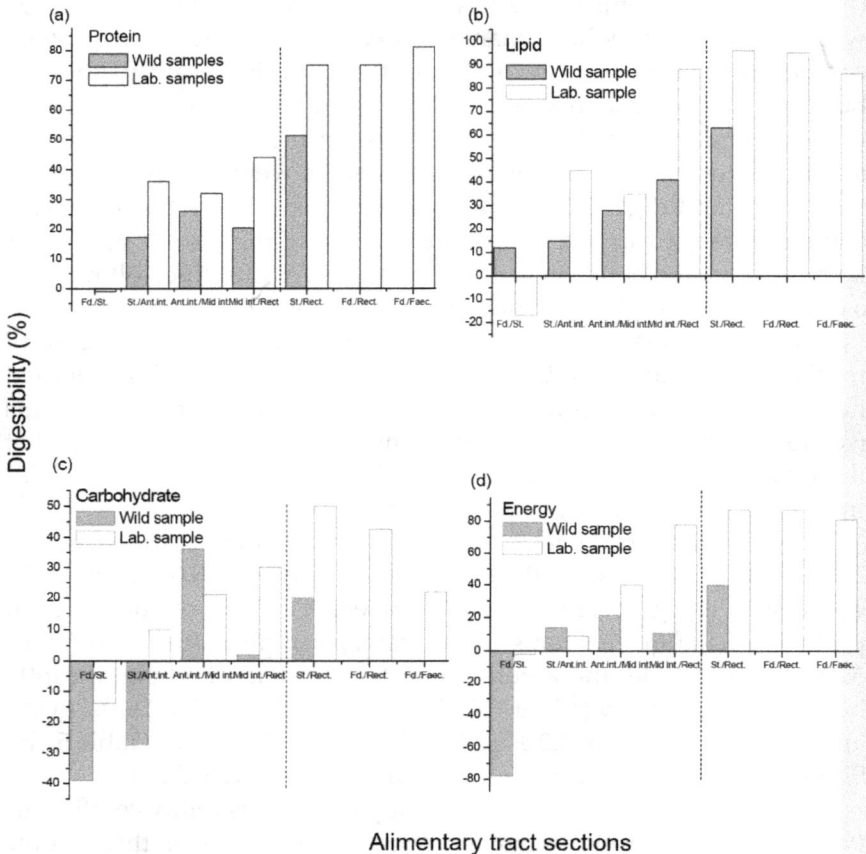

Figure 5.6. Apparent digestibility coefficient for (a) protein, (b) lipid, (c) carbohydrate, and (d) energy between adjacent regions of the archer fish alimentary tract. Gray bars are values for wild archer fish whilst white bars are for laboratory held archer fish. Columns to the right of the vertical dashed line are values for non-adjacent gut sections. For the food/stomach estimate of wild archer fish, the fish were assumed to have eaten mostly crustaceans and insects. (Lab.: laboratory reared samples; Fd.: food; St.: stomach (section a); Ant. int.: Anterior intestine with pyloric caeca (section b); Mid int.: Mid intestine (section c); Rect.: rectum posterior intestine/rectum (section d); Faec.: Faeces). Faeces for laboratory-held archer fish may have lost some soluble nutrient by leaching and ADCs may be over-estimated.

absorption of protein, lipid and energy was detected in all parts of the intestine, including the region between anterior and mid-intestine. In both studies, net absorption had begun as food passed from the stomach into the anterior intestine. The pattern for the small content of carbohydrate, however, was again different. Net addition of carbohydrate to gastrointestinal contents occurred in the stomach, which persisted in the anterior intestine for laboratory fish (Figure 5.6c). Between the stomach and rectum in laboratory archer fish fed with mealworm, the overall effect of absorption in the different zones meant that 75.0% of protein, 95.5% of lipids, 50.0% of carbohydrate and 87.0% of energy had been removed. Comparison of food with faeces in these fish suggested that a further small absorption of all nutrients occurred in the rectum. However, it is equally likely that this apparent increase in ADC could have been caused at least in part by leaching of nutrients from the faeces prior to collection.

The archer fish has a well-developed stomach and a large number of pyloric caeca and a moderately long intestine like cod and whiting. Laboratory-held archer fish that were fed with whole mealworm at meal sizes of 8-13% body wet weight (BW) were found by X-radiography (mealworm was labelled with barium sulphate) and gastrectomy observation to have completely evacuated the stomach within 96 hr post feeding. The decrease in the level of water content and increase in ash content during digestion processes as indicated in this study (Table 5.3b) showed that ashes may play an important role as a compacting factor in digestion processes (Conover 1966; Mazlan and Grove 2003).

Gastric emptying or evacuation rates of several other carnivorous fishes have previously been described using the power exponential model, including black-and-yellow rockfish *Sebastes chrysomelas* (Jordan and Gilbert), Atlantic cod *Gadus morhua* L. and Atlantic herring *Clupea harengus* L. (dos Santos and Jobling 1992; Hopkins and Larson 1990) and recently it was described by using linear, power exponential and logistic model for other carnivorous fishes (Bernes and Murie 2008). However, the result from the present study suggests that a simple square root model can be used to describe gastric emptying in archer fish (*T. jaculatrix*) as reported for other carnivorous fishes (Andersen 1998, 1999, 2001; Andersen and Beyer 2005; Mazlan 2001; Seyhan 1994). Gastric emptying seems best described by a power function: $ds/dt = -\rho S^\alpha$ where S is the current weight of stomach contents, ρ and α are the parameters to be estimated. This description was originally developed by Jones (1974) and afterwards used by dos Santos and Jobling (1992); Temming and Andersen (1994); Andersen (1998, 1999); Mazlan (2001) and Temming and Herrmann (2003). Integration of this differential equation as used in the present study produced non-linear emptying curves which fitted the current data sets well. In the present study, the value of α was found to be 0.5 which was in agreement with the previous report for other carnivorous fishes (Andersen 1998, 1999, 2001; Mazlan 2001).

The present study shows that the concentrations of protein and lipid as well as energy content decreased along the gut. Similar findings were reported for other carnivorous fishes namely, trout (Austreng 1978; Fernández et al. 1998; Lied et al. 1982) and whiting (Mazlan and Grove 2003). There was no indication that any nutrient was selectively retained in or expelled from the stomach relative to other nutrients during the 96 hr study duration. Although the carbohydrate level in the mealworm diet was low, this component increased in the anterior intestine, presumably reflecting secretion of mucus (Table 5.3b).

For protein, 'the small negative ADC shown in Figure 5.6a between original food and stomach contents is most likely due to release of enzymes from the secretory oxyntic peptic' gland cells of the stomach wall, similar findings were reported by Mazlan and Grove (2003) for whiting.

Around 40.0% of the protein leaving the stomach had been absorbed in the short anterior intestine which carries the numerous pyloric caeca. These results agree with those for rainbow trout (Austreng 1978; Dabrowski et al. 1986; Dabrowski and Dabrowska 1981). Krogdahl et al. (1999) found that about 40.0-50.0% of amino acid absorption took place in the pyloric region of salmon. Between the stomach and rectum, about 75.0% of the protein had been removed but the faecal analysis (80.6%) estimate may include errors caused by nutrient leaching into the water. Protein ADCs are usually in the 82.0-93.0% range, although occasional values as low as 70.0% have been recorded (Table 5.4).

Some lipid was apparently 'added' after food entered the stomach. Contents in distal parts of the guts probably come from different meals because of the prolonged digestion time and may explain these anomalies. After reaching the anterior intestine, 45.0% had been absorbed. Lipid digestion and absorption takes place mainly in the anterior intestine where the pyloric caeca are situated and pancreatic lipases are secreted (Borlongan 1990; Buddington and Doroshov 1986; Fänge and Grove 1979). However, in the present study, a significant amount of lipid absorption was observed between the mid-intestine (35.0%) and rectal sections (88.0%), which is consistent with findings reported for other carnivorous fishes (Ferraris and Ahearn 1984; Hofer

Table 5.4. Comparison of the nutrient apparent digestibility (% DW) in various carnivorous fish species

Species	Diets	Sample acquisitions	Protein	Lipids	Carbohydrate	Gross energy	References
			nutrient				
Clarias gariepinus	Fish meal	Food – faeces	-	-	-	90.0	Hossain 1998
Gadus morhua	Fish meal (herring)	Food – stomach	-	-	-	85-95	dos Santos & Jobling 1988
Gadus morhua	Fish meal (Saithe)	Ant. int.– mid. int.	91.0	82.0	-	-	Lied et al. 1982; Lied & Lambertsen 1985
Hippoglossus hippoglossus	Fish meal	Food – faeces	84.1	97.5	82.7	88.4	Grisdale-Helland & Helland 1998
Anguilla rostrata	Fish meal (herring)	Food – faeces	90.7	-	-	90.3	Tibbetts et al. 2000
Oncorhynchus mykiss	Fish meal (herring)	Food – faeces	92.0	97.0	-	91.0	Cho et al. 1985
Oncorhynchus mykiss	Meat + bone meal	Food – faeces	70.0	-	-	-	Dimes & Haard 1994
Pleuronectes platessa	Cod muscle	Food – faeces	91.0	-	-	-	Cowey et al. 1974
Salmo salar	Fish meal (herring)	Food – faeces	82.0	97.9	-	-	Espe et al. 1999

(Contd.)

Table 5.4. (*Contd.*)

Species	Diets	Sample acquisitions	protein	Lipids	Carbohydrate	Gross Energy	References
Salmo salar	Fish meal	Stomach – ant. int.	-	94.0	-	-	Røsjø et al. 2000
Salmo salar	Fish meal	Food – faeces	82.5	94.0	-	-	Sveier et al. 1999
Salvelinus alpinus	Isoenergetic diet	Food – hindgut	-	88.0	-	-	Olsen et. al. 1998
Scienop ocellatus	Fish meal (menhaden)	Food – faeces	77.0	88.0	-	95.0	Gaylord & Gatlin 1996
Sparus aurata	Brown fish meal	Food – faeces	90.8	-	-	-	Fernández et al. 1998
Sparus aurata	Fish meal	Food – faeces	83.0	86.0	82.0	-	Nengas et al. 1997
Dicentrarchus labrax	Fish meal	Food – faces	93.0	-	-	91.0	Gomes Da Silva & Olivia-Teles 1998
Merlangius merlangus	Fresh sprats	Food – faeces	94.1	96.6	80.5	95.8	Mazlan & Grove 2003
Toxotes jaculatrix	Mealworm	Food – faces	80.6	85.6	22.0	81.4	Present study

1982; Koven et al. 1997; Krogdahl et al. 1999; Smith 1989; Smith and Lovell 1973). The absorption of total lipids between stomach and rectum was about 95.5%, and between food and faeces 85.6%, although the latter figure might be increased by nutrient leaching. Typical ADCs for lipid are in the range 86.0-97.5%, although values as low as 82.0% have been reported (Table 5.4). The patterns for removal of energy between the gut sections closely followed those of the high-energy lipids. More than 85.0% of food energy had been removed on reaching the rectum, and faecal contents suggested a maximal ADC of 81.4%. Typical expected values are in the 85.0-95.0% range (Table 5.4).

Absorption of carbohydrate was poor in comparison with protein and lipid, which probably reflected the low carbohydrate content present in the experimental meal. ADC values indicated that considerable addition of carbohydrate occurred in the stomach as well as in the anterior intestine, probably as mucus (Ferraris and Ahearn 1984; Kapoor et al. 1975). Absorption continued in the posterior regions of the gut, leading to overall ADC values of 50.0% (rectum) or 22.0% (faeces). Typical values are about 82.0% (Table 5.4).

The ADC values obtained here are likely to vary under different conditions. Although the data from wild fish give generally similar patterns to those of laboratory fish, the uncontrolled diet led to anomalies (e.g., negative ADC values in the stomach and anterior intestine), which suggests that the field method is less trustworthy. Also, contradictory results among other similar studies for other carnivorous fishes indicate that many variables such as diet types, meal size, ration energy, water temperature, and type of marker, need to be carefully considered for comparability (Cui et al. 1994; Fernández et al. 1998; Flowerdew & Grove 1979; Grove et al. 1978; Jobling 1980a; Mazlan 2001; Mazlan & Grove 2003; Windell 1978).

Conclusion

In conclusion, this study has provided for the first time the basic information on gastric emptying time, digestion and nutrient absorption along the alimentary tract of archer fish *T. jaculatrix*. This information can, in turn, be used for future management, conservation issues and research for archer fish *T. jaculatrix* or any toxotid in Malaysia and nearby coastal waters.

Reproductive Biology of Archer Fish

To facilitate effective management of coastal fishes, detailed information on their reproductive biology is important. However, published information of archer fish reproductive biology is scarce. Allen et al. (2002) and Pethiyagoda (1991) reported that *Toxotes chatareus* females are highly fecund and release between 20,000 and 150,000 eggs; however, no information on their spawning pattern was provided. To date there are no descriptions of the reproductive biology of *T. chatareus*, *T. jaculatrix* or any toxotid, in Malaysia or elsewhere. Consequently, lack of adequate knowledge on the reproductive biology of *T. chatareus* and *T. jaculatrix* remains an impediment to the formulation of sound management strategies for these fascinating fishes.

Therefore, the aim of this chapter is to examine the reproductive biology of *T. chatareus* and *T. jaculatrix* from Sungai Santi, Johor coastal waters, Malaysia focusing on gonadal development of the females, spawning season, sex ratio and fecundity.

The Approach

In the laboratory, ovaries from every fish were examined macroscopically and features such as colour, texture, shape, and turgidity were recorded. These features were used to assign each fish to a maturity stage. Paired gonads were individually weighed to the nearest 0.01 g gonadosomatic index (GSI) and sex ratio (SR) were calculated using the following formula:

$$SR = \frac{Number\ of\ male}{Number\ of\ female}$$

$$GSI = \frac{Weight\ of\ gonads\ (g)}{Weight\ of\ fish\ (g) - Weight\ of\ gonads\ (g)} \times 100$$

Since ovaries are larger and easier to stage compared to testes, only ovaries were used for histological analysis.

Ovaries from all developmental stages were prepared for histological study. Histological sections were prepared from 50 (25 *T. chatareus* and 25 *T. jaculatrix*) preserved ovary samples. Preserved ovaries were sectioned, dehydrated in a graded series of ethanol, and cleared in benzene using an automatic tissue processor (Leica ASP 300, Germany). After 24 hours, samples were embedded in paraffin (Leica Histo embedder, Germany) and sections were cut at 5 μm thickness on a microtome (Leica Multicut Microtome-LR 2045). Sections were dried at 40°C and stained with haematoxylin followed by eosin counter stain, and examined to identify the gametogenic cell types and to subsequently outline the histological features associated with the maturity stages assigned on macroscopic inspection. At least 100 oocytes were randomly selected from each oocyte stage. The diameter of each oocyte was measured to the nearest 0.001 mm on a compound microscope fitted with differential interference contrast optics (BX51 microscope, Olympus, Japan) and fitted with a digital camera in conjunction with colour view III soft imaging system® (Brook Anco, USA).

Fecundity, considered as the total number of oocytes present in mature gonad (ripe) was estimated from 60 females (20 *T. chatareus* and 40 *T. jaculatrix*). Briefly, after fixing in formalin, gonads were carefully dissected and weighed to the nearest 0.1 g. Sub-samples were taken from three different locations of each ovary; near the base, mid and apical regions and weighed to the nearest 0.01 g, oocytes were loosened from trabeculae and counted, and the total number of oocytes in the ovary was estimated, and is reported here as fecundity (Clavier 1992). Relationships between fecundity, total length (TL), eviscerated weight (EW) and gonad weight (GW) were derived by regression analysis.

Analysis of variance and Tukey's post hoc test were used to compare differences in monthly GSI, and oocyte diameters in different stages of ovarian development. Chi-square (χ^2 test) was performed to look for the association between month and sex for both species throughout the sampling period. Statistical significance was inferred at $p < 0.05$. All statistical analyses were performed using SPSS version 15, MINITAB version 14, and Microcal Origin™ version 6.0 software.

The Outcomes

Samples from a total of 358 archer fish were collected during the study. The number of *T. chatareus* captured numbered 164 of which 112 were male, giving a sex ratio of 2.2. *T. chatareus* males ranged in total length from 8.5 cm to 19.4 cm, while females were larger ranging from 9.3 cm to 22.5 cm. The number of *T. jaculatrix* captured numbered 194 of which 139 were male, giving a sex ration of 2.5. *T. jaculatrix* males ranged in total length from 8.6 to 19.0 cm, while females were again larger ranging from 8.7 cm to 23 cm (Figure 6.1).

In both species at all times of year males predominate, however the SR was not consistent throughout the year (Table 6.1), which might be due to different timing of sexual maturity, habitat partitioning or sampling bias. There were no associations observed between month and sexes (i.e., male and female samples were not equally present in each month) for *T. chatareus* ($\chi^2 = 0.63$, $df = 11$, $p > 0.05$) and *T. jaculatrix* ($\chi^2 = 1.77$, $df = 11$, $p > 0.05$). In contrast, when sex ratio is analysed for different sizes classes, the number of males is greater in small sizes classes, whereas males are comparatively rare in larger sizes classes in both species (Figure 6.1). The smaller individuals in the sample (a female of 10 cm and male of 9.5 cm TL in *T. chatareus* and in *T. jaculatrix* female of 10.5 cm and male of 10 cm TL) were already mature so it was not possible to determine the size of the first sexual maturation.

The ovaries of *T. chatareus* and *T. jaculatrix* are semicircular, bilobed, and symmetrical. The major ovarian developmental stages were the same for both species. The terminology used for categorising oocytes based on their macroscopic and microscopic features were adapted from Wallace and Selman (1981); Selman and Wallace (1989); West (1990); Selman et al. (1993); Schaefer (2001); Solomon and Ramnarine (2007); Mazlan and Rohaya (2008) and Arocha and Bárrios (2009) (Table 6.2).

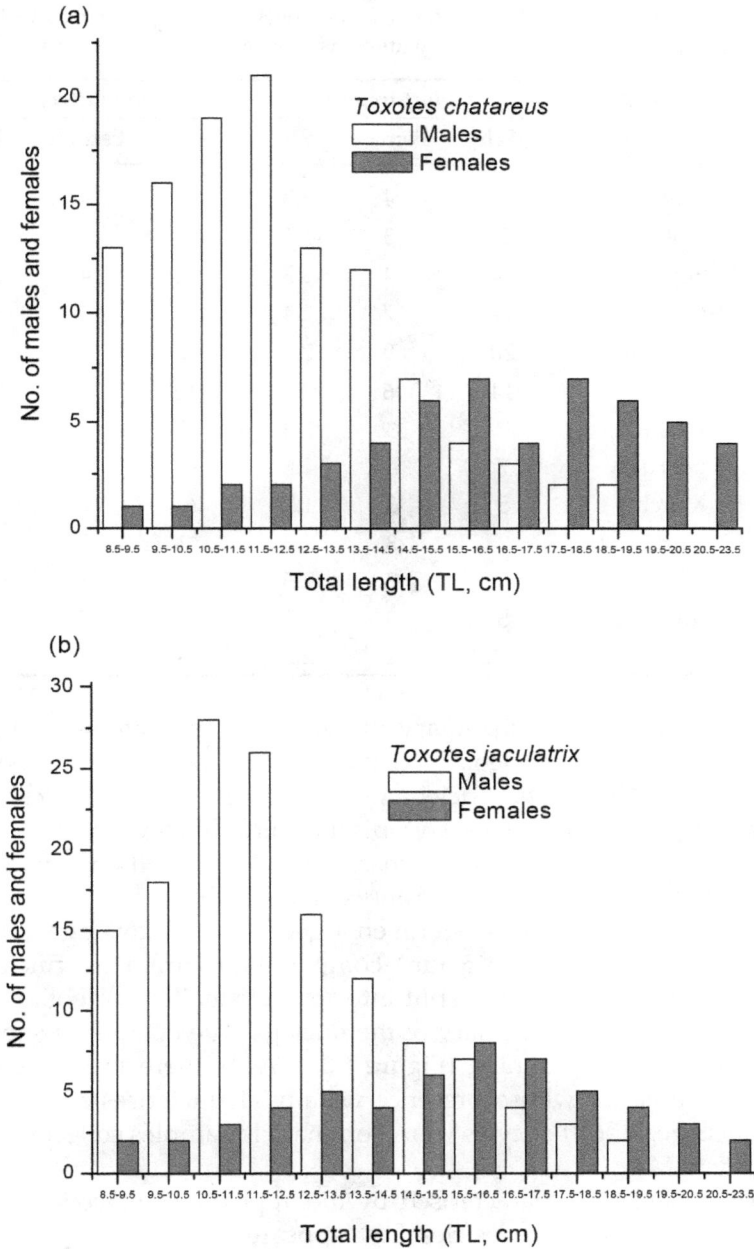

Figure 6.1. Male and female size distribution in the sample (a) *T. chatareus* and (b) *T. jaculatrix* collected from Sungai Santi, Johor coastal waters, Malaysia from July 2008-June 2009.

Table 6.1. Monthly distributions of *T. chatareus* and *T. jaculatrix* collected from Sungai Santi, Johor coastal waters, Malaysia from July 2008-June 2009

Year	Months	T. chatareus			T. jaculatrix		
		Male	Female	SR	Male	Female	SR
2008	July	9	4	2.3	12	4	3.0
	August	5	3	1.7	17	6	2.8
	September	9	4	2.3	10	4	2.5
	October	17	7	2.4	15	5	3.0
	November	20	9	2.2	25	10	2.5
	December	14	6	2.3	12	5	2.4
2009	January	7	3	2.3	7	3	2.3
	February	7	4	1.8	6	4	1.5
	March	5	3	1.7	6	3	2.0
	April	7	3	2.3	12	3	4.0
	May	7	3	2.3	7	3	2.3
	June	5	3	1.7	10	5	2.0
Total		112	52	2.2	139	55	2.5

Maturity stage I or primary growth oocytes, were sub-divided into (i) chromatin nucleolar oocyte and (ii) perinucleolar oocyte (Figure 6.2a). Chromatin nucleolar oocytes were small spherical cells containing a central nucleus. Cytoplasm was a thin layer and strongly basophilic. These oocytes were characterised by small oogonia (40 ± 0.66 µm) and by the presence of only one nucleolus.

Perinucleolus oocytes were enlarged and on average almost double the size (76 ± 0.65 µm) compared to chromatin nucleolar oocytes. The nucleolus had split into several smaller nucleoli, which spread towards the periphery of the nucleus. The cytoplasm showed uniform moderate staining (Figure 6.2a). Also, there were vacuoles in the cytoplasm, whose presence usually characterises the cortical alveoli stage. Stage I oocytes were dominant in samples collected from January to May.

This stage was characterised by the appearance of yolk vesicles in the cytoplasm. The vesicles first appeared as vacuoles but later increased in size and number to form several peripheral rows and gave rise to cortical alveoli, which surrounded the central nucleus (Figure 6.2b). In this phase, the vitellin envelope (zona radiata) began to form

Table 6.2. General macroscopic maturity classification criteria of archer fish ovaries and corresponding microscopic description

Maturity stage	Maturity level	Macroscopic description	Microscopic/histological description	Oocyte diameter (μm)
I	Primary growth (Immature/resting)	Ovaries small, flabby, and pale yellowish in colour.	Ovarian lamellae highly organised. Ovaries contain densely packed unyolked chromatin nucleolar and perinucleolar oocytes. Several vacuoles also present in the cytoplasm (Figure 6.2a)	56 ± 1.5[a]
II	Cortical alveoli (maturing)	Medium size ovary occupying one third to two third length of the body cavity. Ovaries become yellow in colour. Ovaries are firm and turgid in appearance.	Highly organised lamellae. Chromatin nucleolar, perinucleolar and cortical alveolar oocytes present. Contains few to many cortical alveolar in cytoplasm. Cortical alveoli increase in size and number to form several peripheral rows (Figure 6.2b)	90 ± 0.8[b]
III	Vitellogenesis (mature)	Large ovaries occupying two third of the body cavity. Ovaries become orange in colour.	The most advanced group of oocytes in the ovaries contain yolk. Zona radiata, vitellin envelop, follicular epithelium is clear in appearance. Vitellogenic oocytes and cortical alveoli present (Figure 6.2c)	318 ± 2.6[c]

(Contd.)

Table 6.2. (*Contd.*)

Maturity stage	Maturity level	Macroscopic description	Microscopic/histological description	Oocyte diameter (μm)
IV	Spawning	Ovaries have a flabby shrunken appearance and orange or dark yellowish in colour. Remnants of an enlarged blood supply and a thickened ovarian wall. They are distended and occupy most of the body cavity.	Migratory nucleus and hydrated oocytes are the largest group present. Hydrated oocytes appeared irregular or collapsed. (Figure 6.2d)	457 ± 3.1^{d}

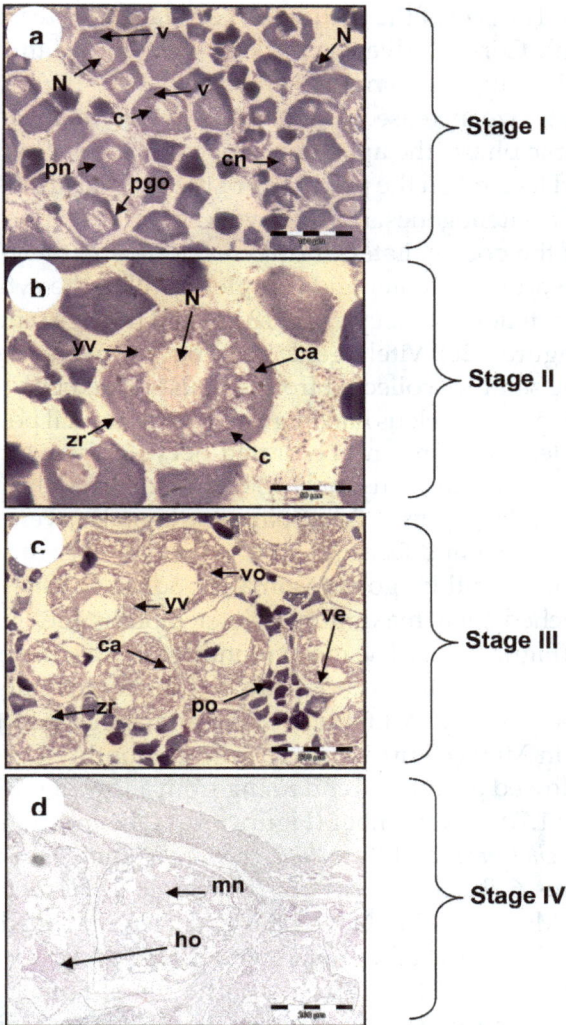

Figure 6.2. Photomicrograph of archer fish ovaries showing different developmental stages of oocyte. (a) Stage I (Primary growth stage or immature or resting): N, nucleus; c, cytoplasm; pgo, primary growth oocyte; pn, perinucleolar oocyte; cn, chromatin nucleolar oocyte; v, vacuoles (40-98 μm). (b) Stage II (Cortical alveoli or maturing): c, cytoplasm; ca, cortical alveoli; yv, yolk vesicle; zr, zona radiata (63-110 μm). (c) Stage III (Vitellogenesis or mature): ca, cortical alveoli; po, primary oocyte; vo, vitellogenic oocyte; ve, vitellin envelop; yv, yolk vesicle; zr, zona radiata (278-389 μm). (d) Stage IV (Spawning): mn, migratory nucleus; ho, hydrated oocytes (378-502 μm). Note that oocyte developmental stages were found to be similar in both species Stage IV (spawning).

(Figure 6.2b). The size of the cortical alveoli stage oocyte was still small (90 ± 0.8 µm). Cortical alveolar oocytes were found throughout the year, except in samples from April and May.

The oocyte size increased dramatically in the vitellogenic phase. In the vitellogenic phase, the appearance of cortical vesicles, which were spherical and located on the periphery of the cytoplasm, was observed. During early vitellogenesis, small yolk granules appeared in the periphery of the oocyte. Later during development, the yolk granules migrated towards the center and completely filled the cytoplasm. The mean size of vitellogenic oocytes is 318 µm and the size ranges from 278 to 389 µm (Figure 6.2c). Vitellogenic stage oocytes were predominately present in the samples collected from August to October.

At this stage the nucleus had migrated to the animal pole. Hydrated mature oocytes were present. Hydrated oocytes always collapsed on histological preparation thus making them look irregular and thereby easily identifiable (Figure 6.2d). Spawning females were found only between November and December.

The mean monthly gonadosomatic indices (GSIs) of male *T. chatareus* reached their maximum in later months (i.e., July, through November) (Figure 6.3a). The mean monthly GSIs for male *T. chatareus* rose from 0.73 in July to reach a peak of 1.51 in November and then declined precipitously to 1.02 in December and reached their lowest GSIs values in May (Figure 6.3a). The mean monthly GSIs of male *T. jaculatrix* followed a similar trend rising from a low of 0.78 in July to a maximum of 1.70 in November (Figure 6.4a). The mean monthly GSIs of female *T. chatareus* and *T. jaculatrix* were higher than those of the males (Figures 6.3b, 6.4b). The GSIs of female *T. chatareus* increased from 2.27 in May to 4.35 in November and then remained at between 3.99 and 2.86 from December to January, before declining in February to May (Figure 6.3b). The mean monthly GSIs of female *T. jaculatrix* showed a similar trend and increased from 2.70 in May to 4.80 in November and then remained between 4.46 and 3.29 from December to January, before declining in February to March (Figure 6.4b).

The majority or all of the females of *T. chatareus* and *T. jaculatrix*, which were caught from July to October, were at the maturing and mature stage except in July, 25% females' ovaries were in immature or resting stage. Females with ovaries at stages IV were only found in November and December, at 80% and 60% respectively. Fish with ovaries at stage I were also recorded in January of the following year and were found until June, and were most abundant in April and

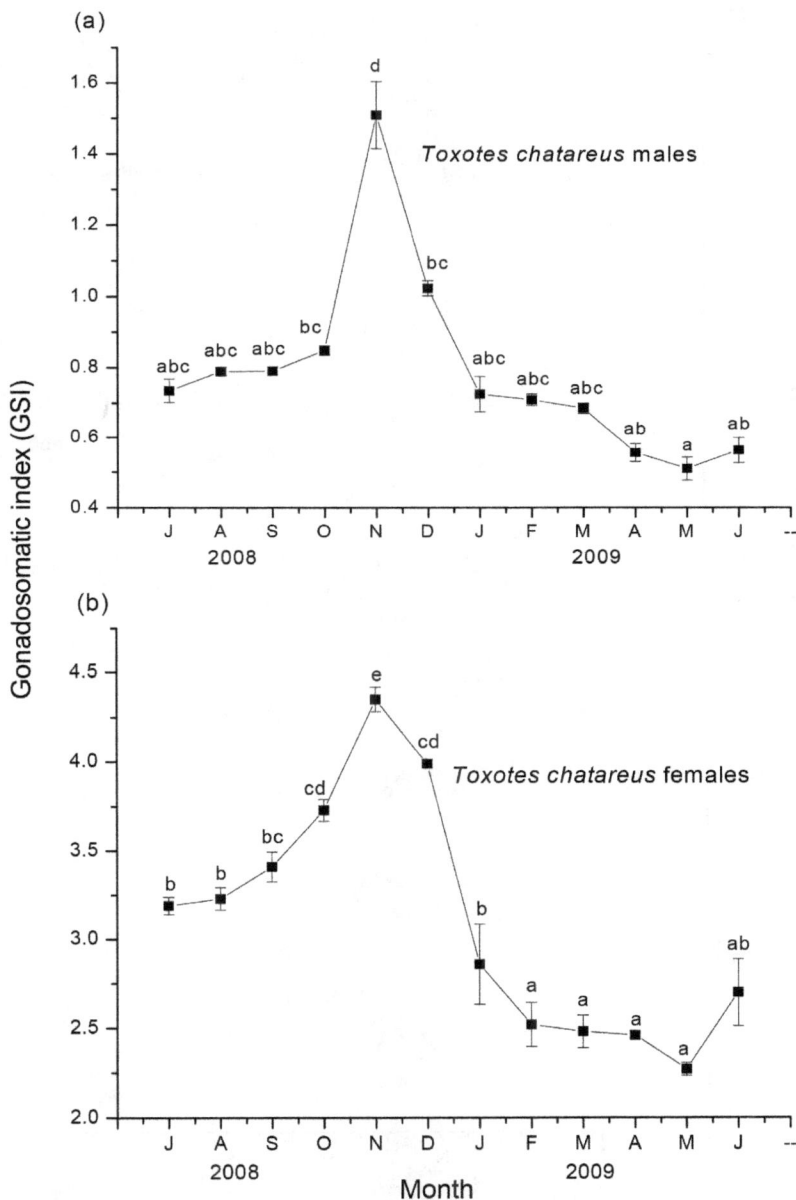

Figure 6.3. Mean monthly gonadosomatic indices ± S.E. of *T. chatareus* (a) males and (b) females, in Sungai Santi, Johor coastal waters, Malaysia. Different letters above the mean values indicate significant differences of mean GSIs at $p < 0.05$.

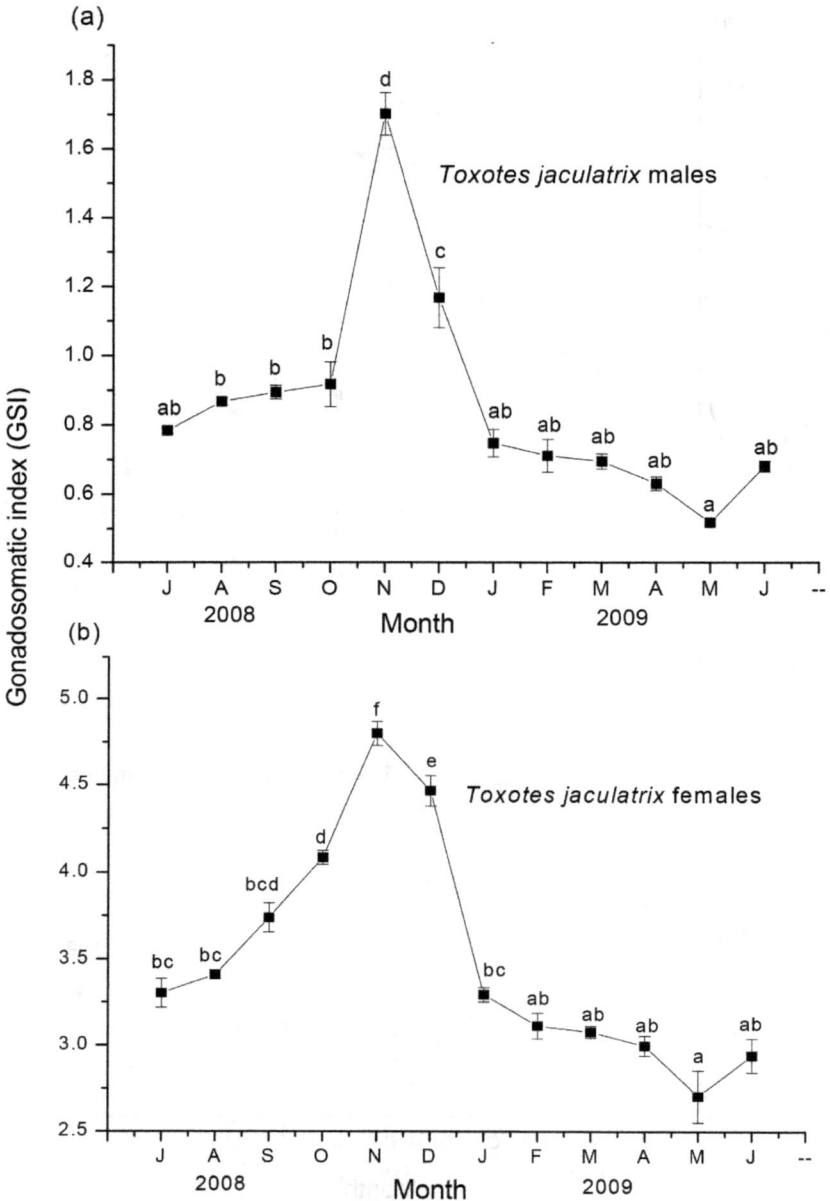

Figure 6.4. Mean monthly gonadosomatic indices ± S.E. of *Toxotes jaculatrix* (a) males and (b) females, in Sungai Santi, Johor coastal waters, Malaysia. Different letters above the mean values indicate significant differences of mean GSIs at $p < 0.05$.

May (Figure 6.5). By June, ovaries were starting to mature and 80% of females were at maturity stage II.

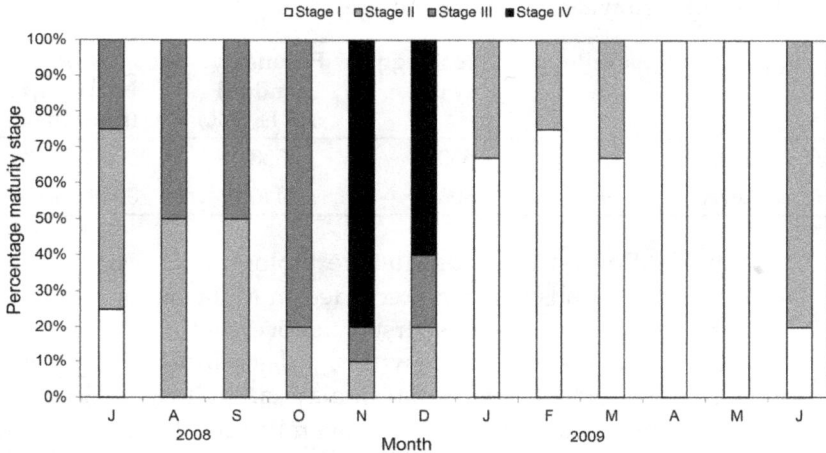

Figure 6.5. Monthly variation in maturity stages of female archer fish (*Toxotes chatareus* and *Toxotes jaculatrix*), in Sungai Santi, Johor coastal waters, Malaysia.

The combined information of the seasonal distribution of mature females, based on ovarian maturation and GSIs results indicated that the spawning season for *T. chatareus* and *T. jaculatrix* occurs from November to December (Figures 6.3-6.5) during the south-east monsoon season (November-December) in Malaysia, which provides the mangrove coastal waters with an abundance of food resources.

Total fecundity 'F' was considered as the number of vitellogenic oocytes at any time in ovary development (Hunter et al. 1992). Total fecundity estimates ranged from 20,000-150,000 eggs for both *T. chatareus* and *T. jaculatrix* females. The sample size, minimum and maximum number of eggs and mean egg counts (S.E.) values are presented in Table 6.3. The relationships of fecundity (F) to body length (TL), body weight (BW), body eviscerated weight (EW), and ovary weight (OW) were examined by correlation analysis (Figure 6.6). There was a strong correlation for each comparison. Results indicate that fecundity increases with total body weight (BW) and eviscerated weight (EW) at a similar rate in both species (Figures 6.6c, d), but with respect to body length, fecundity increases regularly and at a higher rate (Figure 6.6a). However, the highest correlation is found when fecundity is compared with total body weight (BW) ($r = 0.97, 0.93$) and eviscerated weight (EW) (r = 0.97, 0.92) in *T. chatareus* and *T. jaculatrix* respectively (Figures 6.6c, d).

Table 6.3. Estimated fecundity and related statistics in archer fishes

Species	Sample size	Mean egg counts	Fecundity Standard error (± S.E.)	Range: No. of eggs (min.-max.)
T. chatareus	20	55000	5538	20000-150000
T. jaculatrix	40	50000	3440	20000-150000

No information on the reproductive biology of *T. chatareus* and *T. jaculatrix* populations has been recorded in Malaysia or elsewhere. These observations provide the first evidence that *T. chatareus* and *T. jaculatrix* are asynchronous spawners, at least in the Sungai Santi, Johor coastal waters of Peninsular Malaysia. In this geographical location, spawning reaches a peak in November and December. In the study areas both *T. chatareus* and *T. jaculatrix* males dominate in the small size-classes (8.5-15.5 cm TL) with a dramatic inversion in the intermediate sizes (15.5-19.5 cm TL), and larger sizes classes (>19.5 cm TL) are made up of only females. According to Kartas and Quignard (1984), this may be the result of any or all of the following factors: (a) males maturing earlier; (b) a tendency for slower growth in males; and (c) a higher mortality rate in males.

Another possible explanation could be spatial displacement of sexes, which was reported by Cau and Manconi (1983) for other teleosts. However, in the present study male and female samples of both species exhibited similar trends in GSIs throughout the year, suggesting that mature males and females co-occur in the study areas. There was, however, no association observed (p > 0.05) in monthly distribution of male and female samples in the study areas. A possible reason for the apparent abundance of smaller size fish (predominately males) in the total sample could be that larger males are found lower in the water column and are possibly able to evade the nets better.

Archer fish oocytes were classified into four development stages, on the basis of the external features and the histological criteria outlined in Table 6.2 (Selman et al. 1993; Selman and Wallace 1989; Wallace and Selman 1981; West 1990). In archer fish, as in other perciformes (Arocha and Bárrios 2009; Mazlan and Rohaya 2008; Solomon and Ramnarine 2007) oogonia proliferated and turned into primary oocytes, which subsequently grew within follicles, formed cortical alveoli, entered

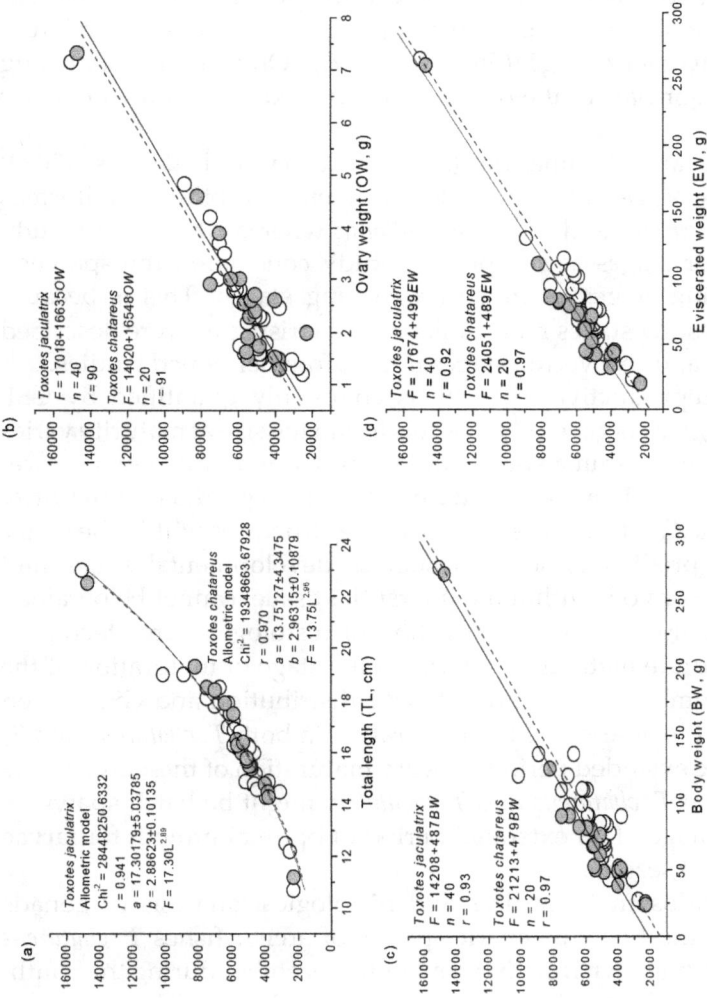

Figure 6.6. Relationship between fecundity and (a) total length (TL); (b) ovary weight (OW); (c) total body weight (BW); and (d) eviscerated weight (EW) in *Toxotes chatareus* and *Toxotes jaculatrix*. (Open circle represent *T. jaculatrix* and filled circle represent *T. chatareus*, dot bar represented non-linear fit of *T. jaculatrix* and solid bar represent non-linear fit of *T. chatareus* in Figure (a), while dot bar represents linear fit of *T. jaculatrix* and solid bar represented linear fit of *T. chatareus* in Figures (b–d)).

vitellogenesis, underwent maturation and were finally ovulated/ spawned. As mentioned before, the preliminary light microscopy observation of gonads showed that various sizes of oocytes were randomly scattered in the gonadal pouch (Table 6.2). This indicated that the oocyte development in archer fish may be asynchronous, meaning that oocytes in all stages of development are present without dominant populations. The ovary appears to be a random mixture of oocytes at every stage (Figures 6.2a-d). Only at the spawning stage, the major part of the ovary was occupied by hydrated oocytes (Figure 6.2d).

Solomon and Ramnarine (2007); Arocha and Bárrios (2009) classified six stages of oocyte development for other perciformes fishes, *Mugil curema* and *Tetrapturus albidus* whereas the present study classified four stages as the present study considered the spawned and recovering stages as immature/resting stages. This is because immature/resting stages had similar characteristics as were described for spawned and recovering stages in the aforementioned studies.

Female reproductive maturity is commonly quantified by GSI, because using size is generally inaccurate and the size at maturity varies greatly not only within a species but even within populations (Lowe-McConnell 1982). However, determination of reproductive maturity using only the GSI is not adequate because structures within the ovary that can be predictive, such as oocytes developmental stage, and accumulation of yolk within the interstitial tissue, cannot be obtained from weight alone. Direct observation of histological architecture is the most accurate method to determine the stage of maturation of the ovary. The combined use of oocyte size distribution and GSI proved to be efficient to assess ovarian maturation in both *T. chatareus* and *T. jaculatrix*. The extended period of ovary maturation of these two spices suggested that *T. chatareus* and *T. jaculatrix* might be batch spawners taking advantage of an extended period of optimal growth for larvae during the wet season.

The gonadosomatic indices and histological analysis of gonads showed that the two congeneric species of archer fishes *T. chatareus* and *T. jaculatrix* spawn from November to December during the Southeast monsoon in Malaysia. Although seasonal changes in temperature and photoperiod are minimal in the tropics, climate is by no means homogenous; seasonal changes in wind and rainfall regimes that affect nutrient inputs do cause seasonality (Lowe-McConnell 1987). Biotic pressure such as competition for spawning grounds or living space may also have an impact (Solomon and Ramnarine 2007). In Johor, Malaysia, the dry season is from January to May and the wet season is from June to December. The timing of archer fish spawning would put the small larvae (hatching size approximately 2.0 mm, Temple unpublished data) in the water column after the first monsoon rains had delivered nutrients from upriver to the estuaries, causing rapid growth of algae and microzooplankton upon which the larvae could feed. Geographical variation in spawning seasons due to local environmental conditions and their effect on the production cycle have been recorded for other perciformes. For example, vermillion snapper (*Rhomboplites aurorubens*) exhibits peak-spawning activity at the beginning of the wet season in the tropics (Trinidad and Tobago) (Manickchand-Heileman and Phillip 1999), whereas peak-spawning activity in the higher latitudes (e.g., Northern Gulf of Mexico) occurs in spring and summer (Nelson 1988). It is suggested that spawning at the higher latitudes is a response to increasing photoperiods in spring which stimulates the spring bloom, whereas peak-spawning activity in tropics (Trinidad) occurs in response to increasing rainfall that results in increased river discharge (Solomon and Ramnarine 2007). Mazlan and Rohaya (2008) also reported that spawning seasonality of giant mudskipper (*Periophthalmodon schlosseri*) was also influenced by south-west monsoon season on the Malaysian coast.

With regards to fecundity, both *T. chatareus and T. jaculatrix* seem to be highly fecund species, when considering that they have freely floating eggs, which are dispersed by the coastal waters from the spawning ground. The results of the present study agree with the findings for *T. chatareus* and *T. jaculatrix* females in previous studies (Allen et al. 2002; Pethiyagoda 1991). The increase in fecundity with body size (length and weight) is consistent with observations made in other species (Barbini and Mc Cleave 1997; Gartner 1993).

Conclusion

In conclusion, this study has described for the first time several key aspects of the reproductive biology of *T. chatareus* and *T. jaculatrix*. The analyses of gonadosomatic index and oocyte diameters, have allowed to define the spawning season, and the estimates of sex ratio and fecundity have implication for resource partitioning in these sympatric species. Histological observations of ovarian development have revealed the basic architecture and stages of oocyte development, and have indicated that both species are most likely asynchronous batch spawners. As this is the first study of the reproductive biology of archer fishes, it provides baseline knowledge for future studies of these fascinating fishes.

References

Abdallah, M. 2002. Length-weight relationship of fishes caught by trawl off Alexandria, Egypt. *Naga ICLARM Quarterly* 25(1): 19-20.

Abdullah, M.J., Karim, S.A. and Hassan, C.H. 2005. Sungai Lebam mangrove forest: A potential ecotourism attraction in Johor. *Forest Biodiversity Series* 4: 380-389.

Adams, S.M. and Breck, J.E. 1990. Bioenergetics. *In:* Schreck, C.B. and Moyle, P.B. (eds.). *Methods for Fish Biology*, pp. 389-409. Bethesda, Maryland: American Fisheries Society.

Adeyemi, S.O., Bankole, N.O., Adikwu, L.A. and Akombu, P.M. 2009. Age, growth and mortality of some commercially important fish species in Gbedikere Lake, Kogi State, Nigeria. *International Journal of Lakes and Rivers* 2(1): 45-51.

Al-Aradi, J.S.J. 1986. Studies on gastric emptying in the turbot, *Scopthalmus maximus* (L.). M.Sc. Thesis, University of Wales Bangor.

Albert, O.T. 1995. Diel changes in food and feeding of small gadoids on a coastal bank. *ICES Journal of Marine Science* 52: 873-885.

Aldman, G. and Larsson, A. 1994. Gastric emptying in the rainbow trout: A gamma scintigraphic study. *In:* Aldman, G. (eds.). *Studies on Gallbladder and Stomach Motility of the Rainbow Trout (Oncorhynchus mykiss), with Special Reference to Cholecystokinin.* PhD Thesis, University of Göteborg.

Allen, G.R. 1978. A review of the archer fishes (family Toxotidae). *Records of Western Australian Museum* 6: 355-378.

Allen, G.R. 1991. *Field Guide to the Freshwater Fishes of New Guinea.* Papua New Guinea: Christensen Research Institute, Madang.

Allen, G.R. 2001. Toxotidae-archer fishes. *In:* Carpenter, K.E. and Niem, V.H. (eds.). FAO Species Identification Guide for Fishery Purposes. *The Living*

Marine Resources of the Western Central Pacific, Vol. 5. Bony Fishes, Part 3 (*Menidaeto Pomacentridae*), Rome: FAO.

Allen, G.R. 2004. *Toxotes kimberleyensis*, a new species of archer fish (Pisces: Toxotidae) from fresh waters of Western Australia. *Records of Western Australian Museum* 56: 225-230.

Allen, G.R., Midgley, S.H. and Allen, M. 2002. *Field Guide to the Freshwater Fishes of Australia*, Perth: Western Australian Museum.

Andersen, N.G. and Beyer, J.E. 2005. Mechanistic modelling of gastric evacuation applying the square root model to describe surface-dependent evacuation in predatory gadoids. *Journal of Fish Biology* 67: 1392-1412.

Andersen, N.G. and Beyer, J. 2008. Predicting ingestion times of individual prey from information about stomach contents of predatory fishes in the field. *Fisheries Research* 92: 1-10.

Andersen, N.G. 1998. Effects of meal size on gastric evacuation in whiting. *Journal of Fish Biology* 52: 743-755.

Andersen, N.G. 1999. The effects of predator size, temperature, and prey characteristics on gastric evacuation in whiting. *Journal of Fish Biology* 54: 287-301.

Andersen, N.G. 2001. A gastric evacuation model for three predatory gadoids and implications of using pooled field data of stomach contents to estimate food rations. *Journal of Fish Biology* 59: 1198-1217.

Anderson, J.S., Lall, S.P., Anderson, D.M. and Chandrasoma, J. 1992. Apparent and true availability of amino acids from common feed ingredients for Atlantic salmon (*Salmo salar*) reared in sea water. *Aquaculture* 108: 111-124.

Andrews, J.W. 1979. Some effects of feeding rate on growth, feed conversion and nutrient absorption of channel catfish. *Aquaculture* 16(3): 243-246.

Arocha, F. and Bárrios, A. 2009. Sex ratios, spawning seasonality, sexual maturity, and fecundity of white marlin (*Tetrapturus albidus*) from the western central Atlantic. *Fisheries Research* 95: 98-111.

Arrhenius, F. 1998. Food intake and seasonal changes in energy content of young Baltic Sea spart (*Sprattus sprattus* L.). *ICES Journal of Marine Science* 55: 319-324.

Austreng, E. 1978. Digestibility determination in fish using chromic oxide marking and analysis of contents from different segments of the gastrointestinal tract. *Aquaculture* 13: 265-272.

Bacheler, N.M., Neal, J.W. and Noble, R.L. 2004. Diet overlap between native bigmouth sleepers (*Gobiomorus dormitor*) and introduced predatory fishes in a Puerto Rico reservoir. *Ecology of Freshwater Fish* 13: 111-118.

Bagenal, T. 1978. *Method for Assessment of Fish Production in Fresh Water*. IBP Handbook No. 3. London: Blackwell Scientific Publications.

Bagenal, T.B. and Tesch, F.W. 1978. Age and growth. *In:* Bagenal, T. (ed.). *Methods for Assessment of Fish Production in Fresh Waters*, pp. 101-136. London: Blackwell Scientific Publications.

Bajkov, A.D. 1935. How to estimate daily food consumption of fish under natural conditions. *Transactions of the American Fisheries Society* 65: 288-289.

Barbini, G.P. and McCleave, J.D. 1997. Fecundity of the American *eel Anguilla rostrata* at 45°N in Maine, USA. *Journal of Fish Biology* 51: 840-847.

Barletta, M., Barletta-Bergen, A., Saint-Paul, U. and Hubold, G. 2005. The role of salinity in structuring the fish assemblages in a tropical estuary. *Journal of Fish Biology* 66: 45-72.

Basimi, R.A. and Grove, D.J. 1985a. Gastric emptying rate in *Pleuronectes platessa* L. *Journal of Fish Biology* 26: 545-552.

Basimi, R.A. and Grove, D.J. 1985b. Estimates of daily food intake by an inshore population of *Pleuronectes platessa* L. off eastern Anglesey, North Wales. *Journal of Fish Biology* 27: 505-520.

Bauchot, R. and Bauchot, M.L. 1978. Coefficient de condition et indice pond6ral chez les T616ost6ens. *Cybium* 3(4): 3-16.

Beamish, F.W.H. and Thomas, E. 1984. Effects of dietary protein and lipid on nitrogen losses in rainbow trout *Salmo gairdneri*. *Aquaculture* 41: 359-371.

Beamish, R.J. and McFarlane, G.A. 1983. The forgotten requirement for age validation in fisheries biology. *Transaction of the American Fisheries Society* 112: 735-743.

Beamish, R.J. and Mcfarlane, G.A. 1987. Current trends in age determination methodology. *In:* Summerfelt, R.C. and Hall, G.E. (eds.). *Age and Growth of Fish*, pp. 15-42. USA: Iowa State University Press.

Bekoff, M. and Dorr, R. 1976. Predation by 'shooting' in archerfish, *Toxotes jaculatrix*: Accuracy and sequences. *British Psychological Society* 7: 167-168.

Berens, E.J. and Murie, D.J. 2008. Differential digestion and evacuation rates of prey in a worm-temperate grouper, gag *Mycteroperca microlepis* (Goode & Bean). *Journal of Fish Biology* 72: 1406-1426.

Bergot, P., Blacn, J.M. and Escaffre, A.M. 1981. Relationship between number of pyloric caeca and growth in rainbow trout (*Salmo gairdnerri* Richardson). *Aquaculture* 22: 81-96.

Beverton, R.J.H. and Holt, S.J. 1957. On the dynamics of exploited fish populations. *Fishery Investigations* (2 Sea Fish.) 19: 533. Ministry of Agriculture, Fisheries, and Food. Great Britain

Beverton, R.J.H. and Holt, S.J. 1996. *On the Dynamics of Exploited Fish Populations*. London: Chapman and Hall.

Bilecenoglu, M. 2009. Growth and feeding habits of the brown comber, *Serranus hepatus* (Linnaeus, 1758) in Izmir Bay, Aegean Sea. *Acta Adriatica* 50(1): 105-110.

Blaber, S.J.M. 2000. *Tropical Estuarine Fishes: Ecology, Exploitation and Conservation*. Oxford: Blackwell Scientific Publications.

Blackith, R.E. and Reyment, R.A. 1971. *Multivariate Morphometrics*. London: Academic Press.

Boetius, J. 1980. Atlantic *Anguilla*: A presentation of old and new data of total numbers of vertebrae with special reference to the occurrence of *Anguilla rostrata* in Europe. *Dana* 1: 1-28.

Boge, G., Rigal, A. and Peres, G. 1979. A study of intestinal absorption in vivo and in vitro of different concentrations of glycine by the rainbow trout (*Salmo gairdneri* Richardson). *Comparative Biochemistry and Physiology* 62(4): 831-836.

Bolger, T. and Connolly, P.L. 1989. The selection of suitable indices for the measurement and analysis of fish condition. *Journal of Fish Biology* 34: 171-182.

Bolles, K.L. and Begg, G.A. 2000. Distinction between silver hake (*Merluccius bilinearis*) stocks in US waters of the northwest Atlantic based on whole otolith morphometrics. *Fishery Bulletin* 98: 451-462.

Bone, Q. and Marshall, N.B. 1982. *Biology of Fishes*. New York: Chapman & Hall.

Bookstein, F.L. 1991. *Morphometric Tools for Landmark Data: Geometry and Biology*. London: Cambridge University Press.

Booth, A.J. and Buxton, C.D. 1997. Management of the panga *Pterogymnus laniarius* (Pisces: Sparidae), on the Agulhas Bank, South Africa using per-recruit models. *Fisheries Research* 32: 1-11.

Borg, B. and Veen, T. Van. 1982. Seasonal effects of photoperiod and temperature on the ovary of the three-spined stickleback, *Gasterosteus aculeatus* L. *Canadian Journal of Zoology* 60: 3387-3393.

Borlongan, I.G. 1990. Studies on the digestive lipases of milkfish, *Chanos chanos*. *Aquaculture* 89: 315-325.

Boyce, S.J., Murray, W.A. and Peck, L.S. 2000. Digestion, gut passage time and absorption efficiency in the Antarctic spiny plunder fish. *Journal of Fish Biology* 57: 908-929.

Braekevelt, C.R. and McMillan, D.B. 1967. Cyclic changes in the ovary of the brook stickelback, *Eucalia inconstans* (Kirtland). *Journal of Morphology* 123: 373-396.

Brett, J.R. and Higgs, D.A. 1970. Effect of temperature on the rate of gastric digestion in fingerling sockeye salmon, *Oncorhynchus nerka*. *Journal of the Fisheries Research Board of Canada* 27: 1767-1779.

Brodeur, R.D. and Pearcy, W.G. 1987. Diel feeding chronology, gastric evacuation and estimated daily ration of juvenile coho salmon, *Oncorhynchus kisutch* (Walbaum), in the coastal marine environment. *Journal of Fish Biology* 31: 465-477.

Brodeur, R.D. 1984. Gastric evacuation rates for two foods in the black rockfish, *Sebastes melanops* Girard. *Journal of Fish Biology* 24: 287-298.

Bromley, P.J. 1994. The role of gastric evacuation experiments in quantifying the feeding rates of predatory fish. *Reviews in Fish Biology and Fisheries* 4: 36-66.

Bromley, P.J., Watson, T. and Hislop, J.R.G. 1997. Diel feeding patterns and the development of food webs in pelagic 0-group cod (*Gadus morhua* L.), haddock (*Melanogrammus aeglefinus* L.), whiting (*Merlangius merlangus* L.), saithe (*Pollachius virens* L.) and Norway pout (*Trisopterus esmarkii* Nilsson) in the northern North Sea. *ICES Journal of Marine Science* 54: 846-853.

Brown, C.A. and Gruber, S.H. 1988. Age assessment of the lemon shark, *Negaprion brevirostris*, using tetracycline validated vertebral centra. *Copeia* 3: 747-753.

Bucke, D. 1971. The anatomy and histology of the alimentary tract of the carnivorous fish pike *Esox lucius* L. *Journal of Fish Biology* 3: 421-431.

Buddington, R.K. and Diamond, J.M. 1987. Pyloric ceca of fish: A "new" absorptive organ. *American Journal of Physiology* 252(1): G65-G76.

Buddington, R.K. and Doroshov, S.I. 1986. Development of digestive secretions in white sturgeon juveniles (*Acipenser transmontanus*). *Comparative Biochemistry and Physiology* 83A: 233-238.

Buddington, R.K., Chen, J.W. and Daimond, J. 1987. Genetic and phenotypic adaptation of intestinal nutrient transport to diet in fish. *Journal of Physiology London* 393: 261-281.

Buddington, R.K., Puchal, A.A., Houpe, K.L. and Diehl, W.I. 1993. Hydrolysis and absorption of two monophosphate derivatives of ascorbic acid by channel catfish *Ictalurus punctatus* intestine. *Aquaculture* 114(3-4): 317-326.

Bull, C.D. and Metcalfe, N.B. 1997. Regulation of hyperphagia in response to varying energy deficits in over wintering juvenile Atlantic salmon. *Journal of Fish Biology* 50: 498-510.

Butler, D.A., Palmer, W.E. and Dowell, S.D. 2004. Passage of arthropod-diagnostic fragments in Northern Bobwhite chicks. *Journal of Field Ornithology* 75: 372-375.

Campana, S.E. and Neilson, J.D. 1982. Daily growth increments in otoliths of starry flounder (*Platichthys stellatus*) and the influence of some environmental variables in their production. *Canadian Journal of Fisheries and Aquatic Sciences* 39: 937-942.

Campana, S.E. and Nelson, J.D. 1985. Microstructure of fish otoliths. *Canadian Journal of Fisheries and Aquatic Sciences* 42: 1014-1032.

Campana, S.E. 1984. Comparison of age determination methods for the starry flounder. *Transactions of the American Fisheries Society* 113: 365-369.

Campana, S.E. 2001. Accuracy, precision and quality control in age determination, including a review of the use and abuse of age validation methods. *Journal of Fish Biology* 59: 197-242.

Campbell, P.J., Houlihan, D.F. and Rennie, M.J. 1997. The use of stable isotopes and the measurement of voluntary food intake in fish physiological studies. First Workshop of COST827, 3-6 April 1997, *Voluntary Food Intake in Fish*.

Casselman, J.M. 1983. Age and growth assessment of fish from their calcified structure-techniques and tools, *NOAA Technical Report NMFS* 8: 1-17.

Casselman, J.M. 1987. Determination of age and growth. *In:* Weatherly, A.H. and Gill, H.S. (eds.). *The Biology of Fish Growth*, London: Academic Press.

Cau, A. and Manconi, P. 1983. Sex-ratio and spatial displacement in *Conger conger* (L.). *Rapp Comm int MerMedit* 28: 93-96.

Cho, C.Y., Cowey, C.B. and Watanabe, T. 1985. *Finfish Nutrition in Asia: Methodological Approaches to Research and Development*, pp. 154. Ottawa, Ontario: IDRC.

Chung, K.C. and Woo, N.Y.S. 1998. Phylogenetic relationships of the Pomacanthidae (Pisces: Teleostei) inferred from allozyme variation. *Journal of Zoology* 246: 215-231.

Chung, K.C. and Woo, N.Y.S. 1999. Age and growth by scale analysis of *Pomacanthus imperator* (Teleostei: Pomacanthidae) from Dongsha Islands, southern China. *Environmental Biology of Fishes* 55: 399-412.

Clavier, J. 1992. Fecundity and optimal sperm density for fertilization in the ormer (*Haliotis tuberculata* L). *In:* Guzman del Proo, S.A., Tegner, M.J. and Shepherd, S.A. (eds.). *Abalone of the World.* Fish Rep Dep Fish 24, Australia: Adelaide.

Claytor, R.R. and MacCrimmon, H.R. 1986. Partitioning size from morphometric data: A comparison of five statistical procedures used in fisheries stock identification research. *Canadian Technical Report of Fisheries and Aquatic Sciences* 1531: 1-31.

Cocheret de la Morinière, E., Nagelkerken, van der Meij, H. and van der Velde, G.I. 2004. What attracts juvenile coral reef fish to mangroves: Habitat complexity or shade? *Marine Biology* 144(1): 139-145.

Conde-padín, P., Graham, J.W. and Rolán-alvarez, E. 2007. Detecting shape differences in species of the *Littorina saxatilis* complex by morphometric analysis. *Journal of Molluscan Studies* 73: 147-154.

Conover, R.J. 1966. Assimilation of organic matter by zooplankton. *Limnology and Oceanography* 11: 338-345.

Cowey, C.B., Adron, J. and Blair, A. 1974. Studies on the nutrition of marine flatfish. Utilization of various dietary proteins by plaice (*Pleuronectes platessa*). *British Journal of Nutrition* 31: 297-306.

Craig, J.F. 1987. *The Biology of Perch and Related Fish.* USA: Timber Press.

Cui, Y., Chen, S. and Shaomei, W. 1994. Effect of ration size on the growth and energy budget of the grass carp *Ctenopharyngodon idella* Val. *Aquaculture* 123: 95-107.

Dabrowski, K. and Dabrowska, H. 1981. Digestion of protein by rainbow trout (*Salmo gairdneri* Rich.) and absorption of amino acids within the alimentary tract. *Comparative Biochemistry and Physiology* 69: 99-111.

Dabrowski, K., Leray, C., Nonnotte, G. and Colin, D.A. 1986. Protein digestion and ion concentrations in rainbow trout (*Salmo gairdneri* Rich.) digestive tract in sea- and fresh-water. *Comparative Biochemistry and Physiology* 83A: 27-39.

Darbyson, E., Swain, D.P., Chabot, D. and Castonguay, M. 2003. Diel variation in feeding rate and prey composition of herring and mackerel in the southern Gulf of St. Lawrence. *Journal of Fish Biology* 63: 1235-1257.

Day, F. 1971. *The Fishes of India.* New Delhi: Today and Tomorrows Book Agency.

de Silva, S.S. and Perera, M.K. 1984. Digestibility in (*Sarotherodon niloticus*) fry: Effect of dietary protein level and salinity with further observations on daily variability in digestibility. *Aquaculture* 38: 293-309.

de Silva, S.S. 1985. Evaluation of the use of internal and external markers in digestibility studies. *In:* Cho, C.Y., Cowey, C.B. and Watanabe, T. (eds.). *Finfish Nutrition in Asia: Methodological Approaches,* pp. 96-102. Ottawa: International Development Research Centre.

Diana, J.S. 1995. *Biology and Ecology of Fishes.* Carmel, Indiana: Cooper Publishing Group LLC.

Dill, L.M. 1977. Refraction and the spitting behaviour of the archer fish (*Toxotes chatareus*). *Behavioural Ecology and Sociobiology* 2: 169-184.

Dimes, L.E. and Haard, N.F. 1994. Estimation of protein digestibility-1: Development of an in vitro method for estimating protein digestibility in salmonids (*Salmo gairdneri*). *Comparative Biochemistry and Physiology* 108: 349-362.

Dorcas, M.E., Peterson, C.R. and Flint, M.E.T. 1997. The thermal biology of digestion in rubber boas (*Charina bottae*): Physiology, behavior, and environmental constraints. *Physiological Zoology* 70: 292-300.

Dorner, H. and Wagner, A. 2003. Size-dependent predator-prey relationships between perch and their fish prey. *Journal of Fish Biology* 62: 1021-1032.

dos Santos, J. and Jobling, M. 1988. Gastric emptying in cod, *Gadus morhua* L.: Effects of food particle size and dietary energy content. *Journal of Fish Biology* 33: 511-516.

dos Santos, J. and Jobling, M. 1990. Aspects of gastric evacuation in the Atlantic cod (*Gadus morhua* L.) *ICES Journal of Coastal and Marine Science* 69: 17.

dos Santos, J. and Jobling, M. 1992. A model to describe gastric evacuation in cod (*Gadus morhua* L.), fed single-meals of natural prey. *ICES Journal of Marine Science* 49: 145-154.

Dryden, I.L. and Mardia, K.V. 1998. *Statistical Shape Analysis*. New York: John Wiley & Sons.

Ecoutin, J.M., Albaret, J.J. and Trape, S. 2005. Length-weight relationships for fish populations of a relatively undisturbed tropical estuary: The Gambia. *Fisheries Research* 72: 347-351.

Edwards, D.J. 1971. Effects of temperature on the rate of passage of food through the alimentary canal of plaice *Pleuronectes platessa* L. *Journal of Fish Biology* 3: 433-439.

Edwards, D.J. 1973. The effect of drugs and nerve section on the rate passage of food through the gut of plaice *Pleuronectes platessa* L. *Journal of Fish Biology* 5: 441-446.

Eggers, D.M. 1977. Factors in interpreting data obtained by diel sampling of fish stomachs. *Journal of the Fisheries Research Board of Canada* 34: 290-294.

Elashoff, J.D., Reedy, T.J. and Meyer, J.H. 1982. Analysis of gastric emptying data. *Gastroenterology* 83: 1306-1312.

Elliott, J.M. and Persson, L. 1978. The estimation of daily rates of food consumption for fish. *Journal of Animal Ecology* 47: 977-991.

Elliott, J.M. 1972. Rates of gastric evacuation in brown trout, *Salmo trutta* L. *Freshwater Biology* 2: 1-18.

Elliott, J.M. 1973. The food of brown and rainbow trout (*Salmo trutta* and *S. gairdneri*) in relation to the abundance of drifting invertebrates in a mountain stream. *Oecologia* 12: 329-347.

Elp, M. and Sen, F. 2009. Biological properties of *Capoeta capoeta* (Guldenstaedt, 1773) population living in Karasu Stream (Van, Turkey). *Journal of Animals and Veterinary Advances* 8(1): 139-142.

Elshoud, G.C.A. and Koomen, P. 1985. A biomechanical analysis of spitting in archerfishes (Pisces, Perciformes, Toxidae). *Zoomorphology* 105: 240-252.

Eschmeyer, W. and Froese, R. 2003. Annotated catalogue of fishes. http://www. calacademy.org/research/ichthyology/annotated/index.html [17 July 2007].

Espe, M., Sveier, H., Hoegoey, I. and Lied, E. 1999. Nutrient absorption and growth of Atlantic salmon (*Salmo salar* L.) fed fish protein concentrate. *Aquaculture* 174: 119-137.

Ezenwa, B.I.O. and Ikusemiju, K. 1981. Age and growth determination in the catfish, *Chrysichthys nigrodigitatus* (Lacepede) by use of the dorsal spine. *Journal of Fish Biology* 19: 345-351.

Fairelough, D. 2005. The biology of four tuskfish species (Choerodon: Labridae) in Western Australia. PhD Thesis, University of Murdoch.

Fänge, R. and Grove, D.J. 1979. Fish digestion. *In:* Hoare, W. (ed.). *Fish Physiology*, pp. 161-260. New York: Academic Press.

Fernández, F., Miquel, A.G., Guinea, J. and Martinez, R. 1998. Digestion and digestibility in Gilthead Sea bream (*Sparatus aurata*): The effect of diet composition and ration size. *Aquaculture* 166(1-2): 67-84.

Ferraris, R.P. and Ahearn, G.A. 1984. Sugar and amino acid transport in fish intestine. *Comparative Biochemistry and Physiology* 77: 397-413.

Ferraris, R.P., Catacutan, M.R., Mabelin, R.L. and Jazul, A.P. 1986. Digestibility in Milkfish, *Chanos chanos* (Forsskal): Effects of protein sources, fish size and salinity. *Aquaculture* 59: 93-105.

Fletcher, D.J. 1984. The physiology control of appetite in fish. *Comparative Biochemistry and Physiology* 78(4): 617-628.

Fletcher, D.J., Grove, D.J., Basimi, R.A. and Ghaddaf, A. 1984. Emptying rates of single and double meals of different food quality from stomach of the dab, *Limanda limanda* (L.). *Journal of Fish Biology* 25: 435-444.

Flowerdew, M.W. and Grove, D.J. 1979. Some observations of the effects of body weight, temperature, meal size and quality on gastric emptying time in the turbot, *Scophthalmus maximus* (L.) using radiography. *Journal of Fish Biology* 14: 229-238.

Forrester, G.E., Chace, J.G. and McCarthy, W. 1994. Diel and density-related changes in food consumption and prey selection by brook char in a New Hampshire stream. *Environmental Biology of Fishes* 39: 301-311.

Fowler, A.J. and Doherty P.J. 1992. Validation of annual growth increments in the otoliths of two species of damselfish from the southern Great Barrier Reef. *Australian Journal of Marine and Freshwater Research* 43: 1057-1068.

Francis, R.I.C.C., Paul, L.J. and Mulligan, K.P. 1992. Ageing of adult snapper (*Pagrus auratus*) from otolith annual ring counts: Validation by tagging and oxytetracycline injection. *Australian Journal of Marine and Freshwater Research* 43: 1069-1089.

Fridriksson, A. 1958. The tribes of North Coast herring of Iceland with special reference to the period 1948-1955. *Rapports du Conseil pourl'Exploration de la Mer* 143: 36-44.

Froese R. and Pauly, D. 2006. FishBase. http://www.fishbase.org [10 January 2008].

Froese, R. 1998. Length-weight relationships for 18 less-studied fish species. *Journal of Applied Ichthyology* 14: 117-118.

Garcia, C.B., Duarte, J.O., Sandoval, N., Von Schiller, D., Melo, G. and Navajas, P. 1998. Length-weight relationships of demersal fishes from the Gulf of Salamanca, Colombia. *Naga ICLARM Quarterly* 21(3): 30-32.

Gartner, J.V. Jr. 1993. Patterns of reproduction in the dominant lantern fish species (Pisces: Myctophidae) of the eastern Gulf of Mexico, with a review of reproduction among tropical-suptropical Myctophidae. *Bulletin of Marine Science* 52(2): 721-750.

Gaylord, T.G. and Gatlin, D.M. 1996. Determination of digestibility coefficients of various feedstuffs for red drum (*Sciaenops ocellatus*). *Aquaculture* 139: 303-314.

Gill, Th. 1909. The archer fish and its feats. *Smithsonian Miscellaneous collections* 52: 277-286.

Godinho, A.L. 1997. Weight-length relationship and condition of the characiform *Triportheus guentheri*. *Environmental Biology of Fishes* 50: 319-330.

Golani, D. and Galil, B. 1991. Trophic relationships of colonizing and indigenous goatfishes (Mullidae) in the eastern Mediterranean with special emphasis on decapod crustaceans. *Hydrobiologia* 218: 27-33.

Gomes, Da Silva, J. and Olivia-Teles, A. 1998. Apparent digestibility coefficients of feedstuffs in seabass (*Dicentrarchus labrax*) juveniles. Aquat. Living Resour./Ressour. *Vivantes Aquatiques* 11: 187-191.

Gonçalves, J.M.S., Bentes, L., Lino, P.G., Ribeiro, J., Canario, A.V.M. and Erzini, K. 1997. Weight-length relationships for selected fish species of the small-scale demersal fisheries of the south and south-west coast of Portugal. *Fisheries Research* 30(3): 253-256.

Gongnet, G.P., Meyer-Burgdorff, K.H., Becker, K. and Guenther, K.D. 1987. The influence of different protein/energy rations and increasing feeding level on nitrogen excretions in growing mirror carp (*Cyprinous carpio* L.). *Journal of Animal Physiology Animal Nutrition* 58(4): 173-188.

Goodbred, S.L., Gilliom, R.J., Gross, T.S., Denslow, N.P., Bryant, W.L. and Schoeb, T.R. 1997. Reconnaissance of 17β-estradiol, 11-ketotestosterone, vitellogenin, and gonadal histopathology in common carp of United States streams: Potential for contaminant-induced endocrine disruption. U.S. Geological Survey Open-File Report 96-627, Sacramento, California, p. 47.

Grisdale-Helland, B. and Helland, S.J. 1998. Macronutrient utilization by Atlantic halibut (*Hippoglossus hippoglossus*): Diet digestibility and growth of 1 kg fish. *Aquaculture* 166: 57-65.

Grove, D.J., Loizides, L.G. and Nott, J. 1978. Satiation amount, frequency of feeding and gastric emptying rate in *Salmo gairdneri*. *Journal of Fish Biology* 12: 507-516.

Grove, D.J., Moctezuma, M.A., Flett, H.R.J., Foott, J.S., Watson, T. and Flowerdew, M.W. 1985. Gastric emptying and the return of appetite in juvenile turbot, *Scophthalmus maximus* L., fed on the artificial diets. *Journal of Fish Biology* 26: 336-354.

Gulland, J.A. 1983. *Fish Population Dynamics: The Implications for Management*, pp. 422. New York: Chichester, John Wiley & Sons Ltd.

Gunn, J.S. and Milward, N.E. 1985. The food, feeding habits and feeding structures of the whiting species, *Sillago sihama* (Forsskal) and *Sillago analis* Whitley from Townsville, North Queensland, Australia. *Journal of Fish Biology* 26: 411-427.

Haddon, M. and Willis, T.J. 1995. Morphometric and meristic comparison of orange roughy (*Haplostethus atlanticus*: Trachichthyidae) from the Puysegur Bank and Lord Howe Rise, New Zealand, and its implications for stock structure. *Marine Biology* 123: 19-27.

Hall, S.J., Gurney, W.S.C., Dobby, H., Basford, D.J., Heaney, S.D. and Robertson, M.R. 1995. Inferring feeding patterns from stomach contents data. *Journal of Animal Ecology* 64: 39-62.

Hanley, F. 1987. The digestibility of food stuffs and the effects of feeding selectivity on digestibility determinations in tilapia, *Oreochromis niloticus*. *Aquaculture* 66: 163-167.

Harmelin-Vivien, M.L., Kaim-Malka, R.A., Ledoyer, M. and Jacob-Abraham, S.S. 1989. Food partitioning among scorpaenid fishes in Mediterranean seagrass beds. *Journal of Fish Biology* 34: 715-734.

Harrison, T.D. 2001. Length-weight relationships of fishes from South African estuaries. *Journal of Applied Ichthyology* 17: 46-48.

Hassan, C.H., Karim, S.A. and Abu, C.R. 2005. Management and conservation of mangroves: Johor experience. *Forest Biodiversity Series* 4: 69-80.

Hastings, W.H. 1969. Nutritional scores. *In:* Neuhaus, O.W. (ed.). *Fish in Research*, pp. 263-293. New York: Academic Press.

Hazel, J.R. 1993. Thermal biology. *In:* Evans, D.H. (ed.). *The Physiology of Fishes*, pp. 427-467. Boca Raton, Florida: CRC Press, Inc.

He, E. and Wurtsbaugh, W.A. 1993. An empirical model of gastric evacuation rates for fish and an analysis of digestion in piscivorous brown trout. *Transactions of the American Fisheries Society* 122: 717-730.

Henken, A.M., Kleingeld, D.W. and Tyssen, P.A.T. 1985. The effects of feeding level on apparent digestibility of dietary dry matter, crude protein and gross energy in the African catfish, *Clarias gariepinus* (Butchell 1822). *Aquaculture* 47: 113-130.

Henriques, L.T., da Silva, J.F.C. and Vasquez, H.M. 2004. Effect of acipin on the degradability and rate of passage of elephant-grass and corn silages in Holstein x Zebu cattle. *Arquivo Brasileiro de Medicina Veterinaria e Zootecnia* 56: 757-763.

Herald, E.S. 1965. How accurate is the archer fish? *Pac Discovery* 9: 12-13.

Hermida, M., Fernández, J.C., Amaro, R. and Miguel, E.S. 2005. Morphometric and meristic variation in Galician threespine stickleback populations, northwest Spain. *Environmental Biology of Fishes* 73: 189-200.

Hernández, F., Takeuchi, T. and Watanabe, T. 1994. Effect of gelatinized corn meal as a carbohydrate source on growth performance, intestinal evacuation, and starch digestion in common carp. *Fisheries Science* 60: 579-582.

Hesp, S.A. Potter, I.C. and Hall, N.G. 2002. Age and size compositions, growth rate, reproductive biology and habitats of West Australian Dhufish, *Glaucosoma herbraicum*, and their relevance to the management of this species. *Fishery Bulletin* 100: 214-227.

Hill, B.J. 1990. Keynote address: Minimum legal sizes and their use in management of Australian fisheries. *In:* D.A. Hancock (ed.). *Legal Sizes and their Uses in Fisheries Management.* Australian Society for Fish Biology Workshop, Lorne, 24 August 1990. Bureau of Rural Resources Proceedings No. 13. Australian Government Publishing Service, Canberra.

Hirji, K.N. 1983. Observation on the histology and histochemistry of the oesophagus of the perch, *Perca fluviatilis* L. *Journal of Fish Biology* 22(2): 145-152.

Hislop, J.R.G., Harris, M.P. and Smith, J.G.M. 1991. Variations in the caloric value and total energy content of the lesser sandeel (*Ammodytes marinus*) and other fish preyed on by seabirds. *Journal of Zoology* 224: 501-517.

Hofer, R. 1982. Protein digestion and proteolytic activity in the digestive tract of an omnivorous cyprinid. *Comparative Biochemistry and Physiology* 72A: 55-63.

Hopkins, T.E. and Larson, R.J. 1990. Gastric evacuation of three food types in the black and yellow rockfish *Sebastes chrysomelas* (Jordan and Gilbert). *Journal of Fish Biology* 36: 673-681.

Hossain, M.A. 1998. Optimization of feeding and growth performance of African catfish (*Clarias gariepinus* Burchell, 1822) fingerlings. Ph.D. Thesis, University of Stirling.

Hossain, M.A.R., Haylor, G.S. and Beveridge, M.C.M. 1998. An evaluation of radiography in studies of gastric evacuation in African catfish fingerlings. *Aquaculture International* 6(5): 379-385.

Hossain, M.A.R., Haylor, G.S. and Beveridge, M.C.M. 2000. The influence of food particles size on gastric emptying and growth rates of fingerling African catfish, *Clarias gariepinus* Burchell 1822. *Aquaculture Nutrition* 6: 73-76.

Htun-Han, M. 1978. The reproductive biology of the dab *Limanda limanda* (L.) in the North Sea: Seasonal changes in the ovary. *Journal of Fish Biology* 13: 351-359.

Hughes, R.N. 1997. Diet selection. *In:* Godin, J.G.J. (ed.). *Behavioural Ecology of Teleost Fishes*, pp. 134-162. London: Oxford University Press.

Hunt, B.P. 1960. Digestion rate and food consumption of Florida gar, warmouth, and largemouth bass. *Transactions of the American Fisheries Society* 89: 206-211.

Hunt, J.N. and MacDonald, I. 1954. The influence of gastric emptying. *Journal of Physiology* 126: 459-474.

Hunter, J.R., Macewicz, B.J., Chyan-huei, L.O.N. and Kimbrell, C.A. 1992. Fecundity, spawning, and maturity of female Dover sole *Microstomus pacificus*, with an evaluation of assumptions and precision. *Fishery Bulletin National Oceanic and Atmospheric Administration*, Washington, D.C. 90: 101-128.

Hyndes G.A., Loneragan, N.R. and Potter, I.C. 1992. Influence of sectioning otoliths on marginal increment trends and age and growth estimates for the flathead *Platycephalus speculator*. *Fishery Bulletin* 90: 276-284.

Hyndes, G.A., Platell, M.E., Potter, I.C. and Lenanton, R.C.J. 1998. Age composition, growth, reproductive biology, and recruitment of King George Whiting, *Sillaginodes punctata*, in coastal waters of southwestern Australia. *Fishery Bulletin* 96: 258-270.

Hynes, H.B.N. 1950. The food of freshwater sticklebacks (*Gasterosteus aculeatus* and *Pygosteus pungitius*) with a review of methods used in studies of the food of fishes. *Journal of Animal Ecology* 19: 36-58.

Hyslop, E.J. 1980. Stomach contents analysis – A review of methods and their application. *Journal of Fish Biology* 17: 411-429.

Ihssen, P.E., Booke, H.E., Casselman, J.M., Mc Glade, J.M., Payne, N.R. and Utter, F.M. 1981. Stock identification: Materials and methods. *Canadian Journal of Fisheries and Aquatic Science* 38: 1838-1855.

Infante, J.L.Z. and Cahu, C.L. 1994. Influence of diet and some pancreatic enzymes in sea bass (*Dicentrachus labrax*) larvae. *Comparative Biochemistry and Physiology* 109A: 209-212.

Innes, S., Lavigne, D.M., Earle, W.M. and Kovacs, K.M. 1987. Feeding rates of seals and whales. *Journal of Animal Ecology* 56: 115-130.

Jennings, S., Kaiser, M.J. and Reynolds, J.D. 2001. *Marine Fisheries Ecology*. Oxford: Blackwell Science.

Jobling, M. 1980a. Gastric evacuation in plaice, *Pleuronectes platessa* L.: Effects of dietary energy level and food composition. *Journal of Fish Biology* 17: 187-196.

Jobling, M. 1980b. Gastric evacuation in plaice, *Pleuronectes platessa* L.: Effects of temperature and fish size. *Journal of Fish Biology* 17: 547-551.

Jobling, M. 1981a. The influence of feeding on metabolic rate of fishes: A short review. *Journal of Fish Biology* 18: 385.

Jobling, M. 1981b. Mathematical models of gastric emptying and the estimation of daily rates of food consumption for fish. *Journal of Fish Biology* 19: 245-257.

Jobling, M. 1981c. Dietary digestibility and influence of food components on gastric evacuation in Plaice, *Pleuronectes platessa* L. *Journal of Fish Biology* 19: 29-36.

Jobling, M. 1983. Effect of feeding frequency on food intake and growth of Arctic charr. *Salvelinus alpinus* (L.). *Journal of Fish Biology* 23: 177-185.

Jobling, M. 1987. Influence of food particle size and dietary energy content on patterns of gastric evacuation in fish: Test of a physiological model of gastric emptying. *Journal of Fish Biology* 30(3): 299-314.

Jobling, M. 1993. Bioenergetics: Feed intake and energy partitioning. *In:* Rankin, J.C. and Jensen, F.B. (eds.). *Fish Ecophysiology*, pp. 1-44. London: Chapman and Hall.

Jobling, M. 1995. *Environmental Biology of Fishes*. London: Chapman and Hall.

Jobling, M. and Davies, P.S. 1979. Gastric evacuation in plaice, *Pleuronestes platessa* L.: Effects of temperature and meal size. *Journal of Fish Biology* 14: 539-546.

Jobling, M., Gwyther, D. and Grove, D.J. 1977. Some effects of temperature, meal size and body weight on gastric evacuation time in the dab *Limanda limanda* (L). *Journal of Fish Biology* 10: 291-298.

Jobling, M., Arnesen, A.M., Baardvik, B.M., Christiansen, J.S. and Jorgensen, E.H. 1995a. Monitoring feeding behaviour and food intake: Methods and applications. *Aquaculture Nutrition* 1(3): 131-143.

Jobling, M., Arnesen, A.M., Baardvik, B.M., Christiansen, J.S. and Jorgensen, E.H. 1995b. Monitoring voluntary feed intake under practical conditions, methods and applications. *Journal of Applied Ichthyology* 11: 248-262.

Johnson, A.G. 1983. Age and growth of yellowtail snapper from South Florida. *Transactions of the American Fisheries Society* 112: 173-177.

Jones, G.P. and McCormick, M.I. 2002. Numerical and energetic processes in the ecology of coral reef fishes. *In:* Sale, P.F. (ed.). *The Ecology of Fishes on Coral Reefs*, San Diego: Academic Press.

Jones, R. 1974. The rate of elimination of food from the stomachs of haddock *Melanogrammus eaeglefinus*, cod, *Gadus morhua* and whiting *Merlangius merlangus*. *Journal du Conseil / Conseil Permanent International pour l'Exploration de la Mer* 35(3): 225-243.

Jorgensen, E. and Jobling, M. 1988. Use of radiographic methods in feeding studies: A cautionary note. *Journal of Fish Biology* 32: 487-488.

Joyce, W.N., Campana, S.E., Natanson, L.J., Kohler, N.E., Pratt, H.L. and Jenson, C.F. 2002. Analysis of stomach contents of the porbeagle shark (*Lamna nasus* Bonnaterre) in the northwest Atlantic. *Journal of Marine Science* 59: 1263-1269.

Kabir, N.M.J., Wee, K.L. and Maguire, G. 1998. Estimation of apparent digestibility coefficients in rainbow trout (*Oncorhynchus mykiss*) using different markers. I: Validation of microtracer F-Ni as a marker. *Aquaculture* 167: 259-272.

Kapoor, B.G., Smit, H. and Verighina, I.A. 1975. The alimentary canal and digestion in teleosts. *Advance Marine Biology* 13: 109-239.

Karakousis, Y., Peios, C., Economidis, P.S. and Triantaphyllidis, C. 1993. Multivariate analyses of the morphological variability among *Barbus peloponnesius* (Cyprinidae) populations from Greece and two populations of *B. meridionalis* and *B. meridionalis petenyi*. *Cybium* 17: 229-240.

Karpouzi, V.S. and Stergiou, K.I. 2003. The relationships between mouth size and shape and body length for 18 species of marine fishes and their trophic implications. *Journal of Fish Biology* 62: 1353-1365.

Kartas, F. and Quignard J.P. 1984. La fécondité des poissons téléostéens. Collection de Biologie des Milieux Marins, Masson, Paris, pp. 19-21.

Katherine, A. 2006. *The Fishes of Kuching Rivers*. Kota Kinabalu: Natural History Publications.

Kimmerer, W., Avent, S.R., Bollens, S.M., Feyrer, F., Grimaldo, L.F., Moyle, P.B., Nobriga, M. and Visintainer, T. 2005. Variability in length-weight relationships used to estimate biomass of estuarine fish from survey data. *Transactions of the American Fisheries Society* 134: 481-495.

Kitchell, J.F. and Windell, J.T. 1968. Rate of gastric digestion in pumpkinseed sunfish, *Lepomis gibbosus*. *Transactions of the American Fisheries Society* 97: 489-492.

Knutsen, I. and Salvanes, A.G. 1999. Temperature-dependent digestion handling time in juvenile cod and possible consequences for prey choice. *Marine Ecological Progress Series* 181: 61-79.

Kohler, C.C. and Ney, J.J. 1982. A comparison of methods for quantitative analysis of feeding selection of fishes. *Environmental Biology of Fishes* 7: 363-368.

Kohler, N.E., Casey, J.G. and Turner, P.A. 1995. Length-weight relationship for 13 species of sharks from the western North Atlantic. *Fishery Bulletin* 93: 412-418.

Kolb, A.R. and Luckey, T.D. 1972. Markers in nutrition. *Nutrition Abstract and Review* 42: 813.

Kolehmainen, S.E. 1974. Daily feeding rates of bluegill (*Lepomis macrochirus*) determined by a refined radioisotope method. *Journal of the Fisheries Research Board of Canada* 31: 67-74.

Koutrakis, E.T. and Tsikliras, A.C. 2003. Length-weight relationships of fishes from three northern Aegean estuarine systems (Greece). *Journal of Applied Ichthyology* 19: 258-260.

Koven, W.M., Henderson, R.J. and Sargent, J.R. 1997. Lipid digestion in turbot (*Scophthalmus maximus*): in-vivo and in-vitro studies of the lipolytic activity in various segments of the digestive tract. *Aquaculture* 151: 155-171.

Krause, J., Jean-Guy, J. and Brown, D. 1998. Body length variation within multi-species fish shoals: The effects of shoal size and number of species. *Oceologia* 114: 67-72.

Krogdahl, A., Nordrum, S., Soerensen, M., Brudeseth, L. and Røsjø, C. 1999. The effects of diet composition on apparent nutrient absorption along the intestinal tract and of subsequent fasting on mucosal disaccharidase activities and plasma nutirent concentration in Atlantic salmon *Salmo salar*. *Aquaculture Nutrition* 5(2): 121-133.

Labropoulou, M., Machias, A., Tsimenides, N. and Eleftheriou, A. 1997. Feeding habits and ontogenetic shift of the striped red mullet, *Mullus surmuletus* Linnaeus, 1758. *Fisheries Research* 31: 257-267.

Labropoulou, M. and Machias, A. 1998. Effect of habitat selection on the dietary patterns of two triglid species. *Marine Ecological Progress Series* 173: 275-288.

Labropoulou, M., Machias, A. and Tsimenides, N. 1999. Habitat selection and diet of juvenile red porgy, *Pagrus pagrus* (Linnaeus, 1758). *Fishery Bulletin* 97: 495-507.

Lachenbruch, P.A. 1967. An almost unbiased method of obtaining confidence intervals for the probability of misclassification in discriminant analysis. *Biometrics* 23: 639-645.

Laegdsgaard, P. and Johnson, C.R. 1995. Mangrove habitats as nurseries: Unique assemblages of juvenile fish in subtropical mangrove in eastern Australia, *Marine Ecology Progress Series* 126: 67-81.

Laegdsgaard, P. and Johnson, C. 2001. Why do juvenile fish utilize mangrove habitats? *Journal of Experimental Marine Biology and Ecology* 257: 229-254.

Langler, K.F. 1956. *Freshwater Fishery Biology*. New York: Wiley & Sons.

Law, A.T., Cheah, S.H. and Ang, K.J. 1985. An evaluation of the apparent digestibility of some locally available plants and a pelleted feed in a three finfish in Malaysia. *In:* Cho, C.Y., Cowey, C.B. and Watanabe, T. (eds.). *Finfish Nutrition in Asia: Methodological Approaches to Research and Development*. Singapore.

Le Cren, E.D. 1951. The length-weight relationship and seasonal cycle in gonad weight and condition in the perch (*Perca fluviatilis*). *Journal of Animal Ecology* 20: 201-219.

Leunda, P.M., Oscoz, J. and Miranda, R. 2006. Length-weight relationships of fishes from tributaries of the Ebro River, Spain. *Journal of Applied Ichthyology* 22: 299-300.

Lied, E., Julshamn, K. and Braekkan, O. 1982. Determination of protein digestibility in Atlantic cod (*Gadus morhua*) with internal and external indicator. *Canadian Journal of Fisheries and Aquaculture Science* 39: 854-861.

Lied, E. and Lambertsen, G. 1985. Apparent availability of fat and individual fatty acids in Atlantic cod (*Gadus morhua*). *Fiskeri direktoratets Skrifter Serie Ernering* 2: 63-75.

Lim, L.H.S. 2006. Diplectanids (Monogenea) on the archerfish *Toxotes jaculatrix* (Pallas) (Toxotidae) off Peninsular Malaysia. *Systematic Parasitology* 64: 13-25.

Lloyd, J.A. and Mounsey, R.P. 1998. The potential of the trammel net as an alternative method for sampling fish on deep water reefs. *Fisheries Research* 39: 67-74.

Lowe-McConnell, R.H. 1982. Tilapias in fish communities. *In:* Pullin, R.S.V. and Lowe-McConnell, R.H. (eds.). *The Biology and Culture of Tilapias*, pp. 83-113. ICLARM Conference Proceedings 7, Phillipines Manila. Proceedings of the International Conference on the Biology and Culture of Tilapias.

Lowe-McConnell, R.H. 1987. *Ecological Studies in Tropical Fish Communities*. London: Cambridge University Press.

Lowerre-Barbieri, S., Chittenden, M.E. Jr. and Jones, C.M. 1994. A comparison of a validated otolith method to age weakfish, *Cynoscion regalis*, with the traditional scale method. *United States Fishery Bulletin* 92: 555-568.

Lu, P.C. 1988. Planar type cast net for fishing. U.S. Patent 4790098, December 13, 1988.

Lugendo, B.R., Nagelkerken, I., van der Velde, G. and Mgaya, Y.D. 2006. The importance of mangroves, mud and sand flats, and seagrass beds as feeding areas for juvenile fishes in Chwaka Bay, Zanzibar: Gut content and stable isotope analyses. *Journal of Fish Biology* 69: 1639-1661.

Lüling, K.H. 1955. Schützenfische (Toxotidae). *Die Aquarien Terrarien* 8(7): 179-184.

Lüling, K.H. 1958. Morphologisch-anatomische und histologische Untersuchungen am Auge des Schützenfisches *Toxotes jaculatrix* (Pallas 1766) (Toxotidae), nebst Bemerkungen zum Spuckgehaben. *Z Morphol Ökol Tiere* 47: 529-610.

Lüling, K.H. 1963. The archerfish. *Scientific American* 209: 100-129.

Lüling, K.H. 1964. Der Schützenfisch *Toxotes jaculatrix* und sein Benehmen. *Acta Societatis Zoologicae Bohemoslovacae* 3: 250-260.

Lüling, K.H. 1969. Spitting at prey by *Toxotes jaculatrix* and *Colisa lalia. Bonner Zoologische Beitraege* 20: 416-422.

Luthy, S.A., Cowen, R.K., Serafy, J.E. and McDowell, J.R. 2005. Toward identification of larval sailfish (*Istiophorus platypterus*), white marlin (*Tetrapturus albidus*), and blue marlin (*Makaira nigricans*) in the western North Atlantic Ocean. *Fishery Bulletin* 103: 588-600.

Machiels, M.A.M. and Henken, A.M. 1985. Growth rate, feed utilization and energy metabolism of the African catfish, *Clarias gariepinus* (Burchell 1822), as affected by the dietary protein and energy content. *Aquaculture* 44: 271-284.

Macpherson, E. 1981. Resource partitioning in a Mediterranean demersal fish community. *Marine Ecological Progress Series* 4: 183-193.

Maeda, M. 2000. Fishing rod. U.S. Patent 6016621, January 25, 2000.

Magnusson, J.J. 1969. Digestion and food consumed by skipjack tuna, *Katsuwonus pelamis. Transactions of the American Fisheries Society* 98: 379-392.

Mamuris, Z., Apostolidis, A.P., Panagiotaki, P., Theodorou, A.J. and Triantaphyllidis, C. 1998. Morphological variation between red mullet populations in Greece. *Journal of Fish Biology* 52: 107-117.

Manickchand-Heileman, S.C. and Phillip, D.A.T. 1999. Contribution to the biology of the vermilion snapper, *Rhomboplites aurorubens*, in Trinidad and Tobago, West Indies. *Environmental Biology of Fishes* 55: 413-421.

Martin-Smith, K.M. 1996. Length-weight relationships of fishes in a diverse tropical fresh-water community, Sabah, Malaysia. *Journal of Fish Biology* 49: 731-734.

Mayekiso, M. and Hecht, T. 1988. Age and growth of *Sandelia bainsii* Castelnau (Pisces: Anabantidae) in the Tyume River, Eastern Cape (South Africa). *South African Journal of Zoology* 23: 295-300.

Maynard, L.A., Loosli, J.K., Hintz, H.F. and Warner, R.G. 1969. *Animal Nutrition*. New York: McGraw-Hill.

Mazlan, A.G. 2001. Food consumption patterns and dietary digestibility of whiting (*Merlangius merlangus* L.) fed in laboratory condition. PhD Thesis, University of Wales Bangor.

Mazlan, A.G. and Grove, D.J. 2003. Gastric digestion and nutrient absorption along the alimentary tract of whiting (*Merlangius merlangus* L.) fed on natural prey. *Journal of Applied Ichthyology* 19: 229-238.

Mazlan, A.G. and Rohaya, M. 2008. Size, growth and reproductive biology of the giant mudskipper, *Periophthalmodon schlosseri* (Pallas, 1770), in Malaysian waters. *Journal of Applied Ichthyology* 24: 290-296.

Meng, H.J. and Stocker, M. 1984. An evaluation of morphometrics and meristics for stock separation of Pacific herring (*Clupea harengus pallasi*). *Canadian Journal of Fisheries and Aquatic Science* 41: 414-422.

Mergardt, N. and Temming, A. 1997. Diel pattern of food intake in whiting (*Merlangius merlangus*) investigated from the weight of partly digested food particles in the stomach and laboratory determined decay functions. *ICES Journal of Marine Science* 54: 226-242.

Milburn, O. and Alexander, R.M.N. 1976. The performance of the muscles involved in spitting by the archerfish *Toxotes*. *Journal of Zoology London* 180: 243-251.

Mills, E.L., Ready, R.C., Jahncke, M., Hanger, C.R. and Trowbridge, C. 1984. A gastric evacuation model for young yellow perch, *Perca flavescens*. *Canadian Journal of Fisheries and Aquatic Sciences* 42: 513-518.

Miranda, R., Oscoz, J., Leunda, P.M. and Escala, M.C. 2006. Weight-length of cyprinid fishes of the Iberian Peninsula. *Journal of Applied Ichthyology* 22: 297-298.

Molnár, G. and Tölg, I. 1960. Roentgenologic investigation of duration of gastric digestion of pike-perch, *Lucioperca lucioperca*. *Acta Biologica Hungary* 11: 103-108.

Molnár, G. and Tölg, I. 1962. Relation between water temperature and gastric digestion of largemouth bass (*Micropterus salmonoides* Lacépéde). *Journal of the Fisheries Research Board of Canada* 19: 1005-1012.

Morales-Nin, B. and Ralston, S. 1990. Age and growth of *Lutzanus kasmira* (Forsskal) in Hawaiian waters. *Journal of Fish Biology* 36: 191-203.

Morey, G., Moranta, J., Massuti, E., Grau, A., Linde, M., Reira, F. and Morales-Nin, B. 2003. Weight-length relationships of littoral to lower slope fishes from the western Mediterrenean. *Fisheries Research* 62: 89-96.

Munro, J.L. and Pauly, D. 1983. A simple method for comparing growth of fishes and invertebrates. *ICLARM Fishbyte* 1(1): 5-6.

Murta, A.G. 2000. Morphological variation of horse mackerel (*Trachurus trachurus*) in the Iberian and North Africa Atlantic: Implications for stock identification. *ICES Journal of Marine Science* 57: 1240-1248.

Myers, G.S. 1952. How the shooting apparatus of the archerfish was discovered. *Aquarius Journal* 23: 210-214.

Nelson, J.S. 1984. *Fishes of the World*, 2nd ed. New York: John Wiley & Sons.

Nelson, J.S. 1994. *Fishes of the World*, 3rd ed. New York: John Wiley & Sons.

Nelson, R.S. 1988. A study of the life history, ecology and population dynamics of four sympatric reef predators (*Rhomboplites aurorubens*, *Lutjanus campechanus*, Lutjanidae, *Haemulon melanurum*, Haemulidae, and *Pagrus pagrus*, Sparidae) on the East and West Flower Garden Banks, Northwestern Gulf of Mexico. PhD Thesis, North Carolina State University.

Nengas, I., Foundoulaki, E., Alexi, M.N. and Papoutsi, E. 1997. Digestibility of nutrients in diets of seabream (*Sparus aurata*) containing different levels of protein and lipids. *Proceedings of the Hellenic Symposium on Oceanography, Fish, and Fisheries, Aquaculture, Inlands Waters* 2: 161-164.

Ney, J.J. 1993. Practical use of biological statistics. *In*: Kohler, C.C. and Hubert, W.A. (eds.). *Inland Fisheries Management in North America*. Bethesda: American Fisheries Society.

Nielson, L.A. and Johnson, D.L. 1983. *Fisheries Techniques*. Bethesda, Maryland: American Fisheries Society.

Norhayati, A., Nizam, M.S., Juliana, W.A., Shukor, M.N., Fizani, A.F.F., Hafizah, S., Sariah, H., Jamaliah, J., Nordira, A.K.I. and Latiff, A. 2005. Comparison of mangrove tree species composition at selected locations in Peninsular Malaysia. *Forest Biodiversity Series* 4: 253-262.

Norman, J.R. and Greenwood, P.H. 1975. *A History of Fishes*. London: Ernest Benn.

O'Dor, R.K. 2003. The unknown ocean. Baseline Report of the Census of Marine Life Research Program. Consortium for Oceanographic Research and Education, Washington DC, pp. 28.

Olsen, R.E., Henderson, R.J. and Ringo, E. 1998. The digestion and selective absorption of dietary fatty acids in Arctic charr, *Salvelinus alpinus*. *Aquaculture Nutrition* 4: 13-21.

Oni, S.K., Olayemi, J.Y. and Adegboye, J.D. 1983. Comparative physiology of three ecologically distinct fresh water fishes, *Alestes nurse* Ruppell, *Synodontis schall* Bloch and *S. schneider* and *Tilapia zilli* Gervais. *Journal of Fish Biology* 22: 105-109.

Oscoz, J., Campos, F. and Escala, M.C. 2005. Weight-length relationships of some fish species of the Iberian Peninsula. *Journal of Applied Ichthyology* 21: 73-74.

Øvredal, J.T. and Totland, B. 2002. The scantrol fish meter for recording fish length, weight and biological data. *Fisheries Research* 55: 325-328.

Özaydin, O. and Taskavak, E. 2007. Length-weight relationships for 47 fish species from Izmir Bay (eastern Agean Sea, Turkey). *Acta Adriatica* 47(2): 211-216.

Pandey, H.S. and Singh, R.P. 1980. Protein digestibility by khosti fish, *Colisa fasciatus* (Pieces, Anabantidae) under the influence of certain factors. *Acta Hydrochimica et Hydrobiologica* 8: 583-585.

Pantulu, V.R. 1961. On the use of pectoral spines for the determination of age and growth of *Mystus gulio* (H). *Proceeding National Institute of Science India* 27B: 1-30.

Paul, L.J. 1967. Early scale growth characteristics of the New Zealand Snapper, *Chrysophrys aurats* (Forster), with reference to selection of a scale-sampling site. *New Zealand Journal of Marine and Freshwater Research* 2: 273-292.

Pauly, D., 1984. Fish population dynamics in tropical waters: A manual for use with programmable calculators. *ICLARM Studies and Reviews* 8: 52-80.

Pauly, D. and Munro, J.L. 1984. Once more on the comparison of growth in fish and invertebrates. *ICLARM Fishbyte* 2(1): 21.

Pauly, D. and Morgan, G.R. 1987. Length-based methods in fisheries research. *ICLARM Conference Proceeding* 13: 468.

Pauly, D. and Christensen, V. 1995. Primary production required to sustain global fisheries. *Nature* 374: 255-257.

Pauly, D., Trites, A., Capuli, E. and Christensen, V. 1995. Diet composition and trophic levels of marine mammals. *ICES Journal of Coastal and Marine Science* 1995/N: 13.

Pauly, D., Trites, A., Capuli, E. and Christensen, V. 1998. Diet composition and trophic levels of marine mammals. *ICES Journal of Marine Science*, 55(3): 467-481.

Pauly, D. and Christensen, V. 2000. Trophic levels of fishes. *In:* Froese, R. and Pauly, D. (eds.). *FishBase 2000: Concepts, Design and Data Sources*, p. 181. ICLARM, Manila, Philippines.

Pauly, D., Froese, R., Sa-a, P., Palomares, M.L., Christensen, V. and Rius, J. 2000. *TrophLab Manual*. ICLARM, Manila.

Pauly, D. and Palomares, M.L. 2000. Approaches for dealing with three sources of bias when studying the fishing down marine food web phenomenon. *In:* Briand, F. (ed.). *Fishing Down the Mediterranean Food Webs?* Vol. 12, pp. 61-66. CIESM Workshop Series.

Pauly, D. 2000a. Predator-prey ratios in fishes. *In:* Froese, R. and Pauly, D. (eds.). *FishBase 2000: Concepts, Design and Data Sources*, p. 201. ICLARM, Manila, Philippines.

Pauly, D. 2000b. Herbivory as a low-latitude phenomenon. *In:* Froese, R. and Pauly, D. (eds.). *FishBase 2000: Concepts, Design and Data Sources*, p. 179. ICLARM, Manila, Philippines.

Pauly, D., Palomares, M.L.D., Froese, R., Sa-a, P., Vakily, M., Preikshot, D. and Wallace, S. 2001. Fishing down Canadian aquatic food webs. *Canadian Journal of Fisheries and Aquatic Sciences* 58: 51-62.

Payne, A.I. and Collinson, R.I. 1983. A comparison of the biological characteristics of *Saotherodon niloticus* (L.) with those of *S. aureus* (Steindachner) and other tilapias of thee delta and lower Nile. *Aquaculture* 30: 335-351.

Peters, D.S. and Hoss, D.E. 1974. A radioisotopic method of measuring food evacuation time in fish. *Transactions of the American Fisheries Society* 103(3): 626-629.

Pethiyagoda, R. 1991. *Freshwater Fishes of Sri Lanka*. The Wildlife Heritage Trust of Sri Lanka, Colombo.

Petrakis, G. and Stergiou, K.I. 1995. Weight-length relationships for 33 fish species in Greek waters. *Fisheries Research* 21: 465-469.

Pierce, G.J., Thorpe, R.S., Hastie, L.C., Brierley, A.S., Boyle, P.R., Guerra, A., Jamieson, R. and Avila, P. 1994. Geographic variation in *Loligo forbesi* in the Northeast Atlantic: Analysis of morphometric data and tests of causal hypotheses. *Marine Biology* 119: 541-547.

Pillay, T.V.R. 1952. A critique of the methods of study of food of fishes. *Journal of the Zoological Society of India* 4: 185-200.

Pimm, S.L. and Lawton, J.H. 1978. On feeding on more than one trophic level. *Nature* 275: 542-544.

Pitcher, T.J. and Hart, P.J. 1982. *Fisheries Ecology*, London: Chapman and Hall.

Pongsuwana, U., Rahman, A.A. and Khamis, R. 1984. Malaysia Coastal Aquaculture Development. FAO Field document.

Post, D.M., Conners, M.E. and Goldberg, D.S. 2000. Prey preference by a top predator and the stability of linked food chains. *Ecology* 81: 8-14.

Pouyaud, L., Teugels, G.G. and Legendre, M. 1999. Description of a new pangasiid catfish from south-east Asia (Siluriformes, pangasiidae). *Cybium* 23: 247-258.

Punt, A.E. and Hughes, G.S. 1992. PC-YIELD II User's Guide. Bengwela Ecology Programme Report No. 26, Foundation for Research and Development, South Africa.

Pütter, A. 1920. Studien über physiologische Ähnlichkeit. VI. Wachstumsähnlichkeiten. *Pflüger Archiv für die gesamte Physiologie des Menschen und der Tiere* 180: 298-340.

Reimchen, T.E., Stinson, E.M. and Nelson, J.S. 1985. Multivariate differentiation of parapatric and allopatric populations of threespine stickleback in the Sangan River watershed, Queen Charlotte Islands. *Canadian Journal of Zoology* 63: 2944-2951.

Riche, M., Haley, D.I., Oetker, M., Garbrecht, S. and Garling, D.L. 2004. Eject of feeding frequency on gastric evacuation and the return of appetite in tilapia *Oreochromis niloticus* L. *Aquaculture* 234: 657-673.

Ricker, W.E. 1973. Linear regressions in fishery research. *Journal of the Fisheries Research Board of Canada* 30: 409-434.

Ricker, W.E. 1975. Computation and interpretation of biological statistics of fish populations. *Bulletin of the Fisheries Research Board Canada* 191: 1-382.

Rideout, R.M., Maddock, D.M. and Burton, M.P.M. 1999. Oogenesis and the spawning pattern in Greenland halibut from the North-west Atlantic. *Journal of Fish Biology* 54: 196-207.

Rindorf, A. 2002. The effect of stomach fullness on food intake of whiting in the North Sea. *Journal of Fish Biology* 61: 579-593.

Robb, A.P. 1990. Gastric evacuation in whiting (*Merlangius merlangius* L.). ICES CM/G:51 Demersal Fish Committee Session 0, pp. 5.

Robertson, A.I., Alongi, D.M. and Boto, K. 1992. Food chains and carbon fluxes. *In:* Robertson, A.I. and Alongi, D.M. (eds.). *Tropical Mangrove Ecosystems: Coastal and Estuarine Studies*, pp. 293-326. American Geophysical Union, Washington.

Rönnbäck, P., Troell, M., Kautsky, N. and Primavera, J.H. 1999. Distribution pattern of shrimps and fish among *Avicennia* and *Rhizophora* microhabitats in the Pagbilao Mangroves, Philippines. *Estuarine Coastal and Shelf Science* 48: 223-234.

Rønnestad, I., Rojas-Garcia, C.R. and Skadal, J. 2000. Retrograde peristalsis: A possible mechanism for filling the phyloric caeca? *Journal of Fish Biology* 55(6): 216-218.

Røsjø, C., Nordrum, S., Olli, J.J., Krogdahl, A., Ruyter, B. and Holm, H. 2000. Lipid digestibility and metabolism in Atlantic salmon (*Salmo salar*) fed medium-chain triglycerides. *Aquaculture* 190: 65-76.

Ross, B. and Jauncey, K. 1981. A radiographic estimation of the effect of temperature on gastric emptying in *Sarotherodon niloticus* (L.), *S. aureus* (Steindachner) hybrids. *Journal of Fish Biology* 19: 333-344.

Rossel, S., Corlija, J. and Schuster, S. 2002. Predicting three dimensional target motion: How archer fish determine where to catch their dislodged prey. *Journal of Experimental Biology* 205: 3321-3326.

Roxburgh, L. and Pinshow, B. 2002. Digestion of nectar and insects by Palestine sunbirds. *Physiological and Biochemical Zoology* 75: 583-589.

Royce, W.F. 1972. *Introduction to the Fisheries Science.* New York: Academic Press.

Russel, N.R. and Wootton, R.J. 1993. Satiation, digestive-tract evacuation and return of appetite in European Minnow, *Phoxinus phoxinus* (Cyprinidae) following short period of preprandial starvation. *Environmental Biology of Fishes* 38(4): 385-390.

Sadovy, Y. 1996. Reproduction in reef fishery species. *In:* Polunin, N.V.C. and Roberts, C.M. (eds.). *Reef Fisheries*, pp. 15-59. London: Chapman and Hall.

Safran, P. 1992. Theoretical analysis of the weight-length relationships in the juveniles. *Marine Biology* 112: 545-551.

Salvanes, A.G.V., Aksnes, D.L. and Giske, J. 1995. A surface-dependent gastric evacuation model for fish. *Journal of Fish Biology* 47: 679-695.

Samat, A., Shukor, M.N., Mazlan, A.G., Arshad, A. and Fatimah, M.Y. 2008. Length-weight relationship and condition factor of *Pterygoplichthys pardalis* (Pisces: Loricariidae) in Malaysia Peninsula. *Research Journal of Fisheries and Hydrobiology* 3(2): 48-53.

Sarre, G.A. and Potter, I.C. 2000. Variation in age compositions and growth rates of *Acanthopagrus butcheri* (Sparidae) among estuaries: Some possible contributing factors. *Fish Bulletin* 98: 785-799.

Sasekumar, A., Chong, C., Leh, M.U. and D'Cruz, R. 1992. Mangroves as habitat for fish and prawns. *Hydrobiologia* 247: 195-207.

Schaefer, K. 2001. Reproductive biology of tunas. *In:* Block, B. and Stevens, E.D. (eds.). *Tuna Physiology, Ecology, and Evolution. Fish Physiology*, pp. 225-270. New York: Academic Press.

Scharf, F.S., Juanes, F. and Rountree, R.A. 2000. Predator size–prey size relationships of marine fish predators: Interspecific variation and effects of ontogeny and body size on trophic niche breadth. *Marine Ecology Progress Series* 208: 229-248.

Schlosser, J.A. 1764. An account of a fish from Batavia called Jaculator. In a letter to Mr. Peter Collinson, F.R.S. from John Albert Schlosser, M.D.F.R.S. *Philosophical Transactions of the Royal Society London* 54: 89-91.

Schuster, S., Rossel, S., Schmidtmann, A., Jäger, A. and Poralla, J. 2004. Archer fish learn to compensate for complex optical distortions to determine the absolute size of their aerial prey. *Current Biology* 14: 1565-1568.

Schuster, S., Wohl, S. and Griebsch, M. 2006. Animal cognition: How archerfish learn to down rapidly moving targets. *Current Biology* 16: 378-383.

Seaburg, K.G. and Moyle, J.B. 1964. Feeding habits, digestion rates and growth of some Minnesota warm water fishes. *Transactions of the American Fisheries Society* 93: 269-285.

Secor, D.H., Dean, J.M. and Laban, E.H. 1991. Manual for otolith removal and preparation for microstructural examination. Baruch Institute Technical Report, University of South Carolina, Columbia.

Selman, K. and Wallace, R.A. 1989. Cellular aspects of oocyte growth in teleost. *Zoological Science* 6: 211-231.

Selman, K., Wallace, R.A., Sarka, A. and Qi, X. 1993. Stages of oocyte development in the zebrafish, *Brachydanio rerio. Journal of Morphology* 218: 203-224.

Senar, J.C., Leonart, J. and Metcalfe, N.B. 1994. Wing shape variation between resident and transient wintering siskins *Carduelis spinus*. *Journal of Avian Biology* 25: 50-54.

Seyhan, K. 1994. Gastric emptying, food consumption and ecological impact of whiting, *Merlangius merlangus* (L.) in the Irish Sea marine ecosystem. PhD Thesis, University of Wales Bangor.

Seyhan, K. and Grove, D.J. 1998. Food consumption of whiting, *Merlangius merlangus* in the Eastern Irish Sea. *Fisheries Research* 38: 233-245.

Seyhan, K. and Grove, D.J. 2003. A new approach in modeling gastric emptying in fish. *Turkey Journal of Veterinary and Animal Science* 27: 1043-1047.

Shepherd, G.R. 1988. Age determination methods for northwest Atlantic species, weakfish *Cynoscion regalis*. *In:* Penttila, J. and Derry, L.M. (eds.). *Age Determination Methods for Northwest Atlantic Species*, pp. 71-76. NOAA Tech. Rep. NMFS 72.

Sim, D.W., Davies, S.J. and Bone, Q. 1996. Gastric emptying rate and return of appetite in lesser spotted dogfish, *Scyliorhinus canicula* (Chondrichthyes: Elasmobranchii). *Journal of Marine Biology Association U.K.* 76(2): 479-491.

Sin, Y.W., Yau, C. and Chu, K.H. 2009. Morphological and genetic differentiation of two loliginid squids, *Uroteuthis* (*Photololigo*) *chinensis* and Uroteuthis (*Photololigo*) *edulis* (Cephalopoda: Loliginidae), in Asia. *Journal of Experimental Marine Biology and Ecology* 369: 22-30.

Singh-Renton, S. and Bromley, P.J. 1996. Effects of temperature. Prey type and prey size on gastric evacuation in small cod and whiting. *Journal of Fish Biology* 49: 702-713.

Sinovcic, G., Franicevic, M., Zorica, B. and Cikes-Kec, V. 2004. Length-weight and length-length relationships for 10 pelagic fish species from the Adriatic Sea (Croatia). *Journal of Applied Ichthyology* 20: 156-158.

Sis, R.F., Ives, P.J., Jones, D.M., Lewis, D.H. and Haensly, W.E. 1979. The microscopic anatomy of the oesophagus, stomach and intestine of the channel catfish, *Ictalurus punctatus*. *Journal of Fish Biology* 14(2): 179-186.

Six, L.D. and Horton, H.F. 1977. Analysis of age determination methods for yellow tail rockfish, canary rockfish and black rockfish of Oregon. *United States Fishery Bulletin* 75: 405-414.

Smith, B.W. and Lovell, R.T. 1973. Determination of apparent digestibility in feeds for channel catfish. *Transactions of the American Fisheries Society* 4: 831-835.

Smith, H.M. 1945. The fresh-water fishes of Siam, or Thailand. *Bulletin of the United States National Museum* 188: 1-622.

Smith, L.S. 1989. Digestive functions in teleost fishes. *In:* Harver, J.E. (ed.). *Fish Nutrition*. San Diego: Academic Press.

Solomon, F.N. and Ramnarine, I.W. 2007. Reproductive biology of white mullet, *Mugil curema* (Valenciennes) in the Southern Caribbean. *Fisheries Research* 88: 133-138.

Sparre, P. 1992. *Introduction to Tropical Fish Stock Assessment*. Part I. Manual. FAO Fisheries Technical Paper 306/1. Rev 1, Rome, Italy.

Spiegel, M.R., 1991. *Théorie et Applications de la Statistique*. Paris: McGraw-Hill.

Sponheimer, M., Robinson, T. and Roeder, B. 2003. Digestion and passage rates of grass hays by llamas, alpacas, goats, rabbits, and horses. *Small Ruminant Research* 48: 149-154.

Srijunngam, J. and Wattanasirmkit, K. 2001. Histological structure of Nile Tilapia *Oreochromis niloticus* Linn. Ovary. *The Natural History Journal of Chulalongkorn University* 1(1): 53-59.

Stergiou, K.I. and Fourtouni, H. 1991. Food habits, ontogenetic diet shift and selectivity in *Zeus faber* Linnaeus, 1758. *Journal of Fish Biology* 39: 589-603.

Stergiou, K.I. and Karpouzi, V.S. 2002. Feeding habits and trophic levels of Mediterranean fish. *Review in Fish Biology and Fisheries* 11: 217-254.

Stergiou, K.I. 1988. Feeding habits of the Lessepsian migrant *Siganus luridus* in the eastern Mediterranean, its new environment. *Journal of Fish Biology* 33: 561-543.

Stevens, C.E. and Hume, I.D. 1995. *Comparative Physiology of the Vertebrate Digestive System*, 2nd ed. Cambridge, London: Cambridge University Press.

Stevenson, D.K. and Campana, S.E. 1992. Otolith microstructure examination and analysis. *Canadian Special Publication of Fisheries and Aquatic Sciences* 117: 126.

Storebakken, T., Austreng, E. and Steenberg, K. 1981. A method for determination of food intake in salmonics using radioactive isotopes. *Aquaculture* 24: 133-142.

Sullivan, S.O., Moriarty, C., Fitzgerald, R.D., Davenport, J. and Mulcahy, M.F. 2003. Age, growth and reproductive status of the European conger eel, *Conger conger* (L.) in Irish coastal waters. *Fisheries Research* 64: 55-69.

Sveier, H., Wathne, E. and Lied, E. 1999. Growth, feed and nutrient utilization and gastrointestinal evacuation time in Atlantic salmon (*Salmo salar*): The effect of dietary fish meal particle size and protein concentration. *Aquaculture* 180(3-4): 265-282.

Sweka, J.A., Cox, M.K. and Hartman, K. 2004. Gastric evacuation rates of Brook trout. *Transaction of the American Fisheries Society* 133: 204-210.

Swenson, W.A. and Smith, L.L. 1973. Gastric digestion, food conversion, feeding periodicity, and food conversion efficiency in walleye (*Stizostedion vitreum*). *Journal of the Fisheries Research Board of Canada* 30: 1327-1336.

Talbot, C. and Higgins, P.J. 1983. A radiographic method for feeding studies on fish using metallic iron powder as a marker. *Journal of Fish Biology* 23: 211-220.

Talbot, C. 1985. Laboratory methods in fish feeding and nutritional studies. *In:* Tytler, P. and Calow, P. (eds.). *Fish Energetics New Perspectives*, pp. 125-154. London and Sydney: Croom Helm.

Taskavak, E. and Bilecenoglu, M. 2001. Length weight relationships for 18 Lessepsian (Red Sea) immigrant fish species from the eastern Mediterranean coast of Turkey. *Journal of the Marine Biological Association of the United Kingdom*, 81(5): 895-896.

Temming, A. and Andersen, N.G. 1994. Modeling gastric evacuation without meal size as a variable. A model applicable for the estimation of daily ration of cod (*Gadus morhua* L.) in the field. *ICES Journal of Marine Science* 51: 429-438.

Temming, A. and Herrmann, J.-P. 2003. Gastric evacuation in cod Prey-specific evacuation rates for use in North Sea, Baltic Sea and Barents Sea multi-species models. *Fisheries Research* 63: 21-41.

Temple, S.E. 2007. Effect of salinity on the refractive index of water: Consideration for archer fish aerial vision. *Journal of Fish Biology* 70: 1626-1629.

Thorpe, J.E. 1977. Daily ration of adult perch, *Perca fluviatilis* L., during summer in Loch Leven, Scotland. *Journal of Fish Biology* 11: 55-68.

Thorpe, R.S. 1976. Biometric analysis of geographic variation and racial affinities. *Biological Reviews* 51: 407-452.

Thorrson, J., Mayes, R.W. and Gudmundsson, O. 1997. Incorporation of marker n-alkanes into experimental feed. First Workshop of COST827, Voluntary Food Intake in Fish, 3-6 April 1997.

Tibbetts, S.M., Lall, S.P. and Anderson, D.M. 2000. Dietary protein requirement of juvenile American eel (*Anguilla rostrata*) fed practical diets. *Aquaculture* 186: 145-155.

Timmermans, P.J.A. 1975. The prey catching behaviour of the archerfish *Toxotes*. *Netherlands Journal of Zoology* 25: 381.

Timmermans, P.J.A. 2000. Prey catching in the archerfish: Marksmanship and endurance of squirting at an aerial target. *Netherlands Journal of Zoology* 50: 411-423.

Timmermans, P.J.A. and Maris, E. 2000. Does the bright spot on the back of young archer fishes serve group coherence? *Netherlands Journal of Zoology* 50: 4.

Timmermans, P.J.A. and Vossen, J.H.M. 2000. Prey catching in the archerfish: Does the fish use a learned correction for refraction? *Behavioral Processes* 52: 21-34.

Timmermans, P.J.A. 2001. Prey catching in the archerfish: Angles and probability of hitting an aerial target. *Behavioral Processes* 55: 93-105.

Timmermans, P.J.A. and Souren, P.M. 2004. Prey catching in archer fish: The role of posture and morphology in aiming behavior. *Physiology and Behavior* 81: 101-110.

Titus, K., Mosher, J.A. and Williams, B.K. 1984. Chance-corrected classification for use in discriminant analysis-ecological applications. *American Midland Naturalist* 111: 1-7.

Treasurer, J.W. and Holliday, F.G.T. 1981. Some aspects of the reproductive biology of perch *Perca fluviatilis* L. A histological description of the reproductive cycle. *Journal of Fish Biology* 18: 359-376.

Tudela, S. 1999. Morphological variability in a Mediterranean, genetically homogeneous population of the European anchovy, *Engraulis encrasicolus*. *Fisheries Research* 42: 229-243.

Turan, C., Oral, M., Öztürk, B. and Düzgüneş, E. 2006. Morphometric and meristic variation between stocks of Bluefish (*Pomatomus saltatrix*) in the Black, Marmara, Aegean and northeastern Mediterranean Seas. *Fisheries Research* 79: 139-147.

Tyler, A.V. 1970. Rate of gastric emptying in young cod. *Journal of the Fisheries Research Board of Canada* 27: 1177-1189.

Ursin, E. 1968. A mathematical model of some aspects of fish growth, respiration and mortality. *Journal of the Fisheries Research Board of Canada* 24: 2355-2453.

Verwey, J. 1928. Iets over de voedingswijze van *Toxotes jaculator*. *De Tropische Natuur* 17: 162-166.

von Bertalanffy, L. 1938. A quantitative theory of organic growth (inquiries on growth laws. II). *Human Biology* 10: 181-213.

Wallace, R. and Selman, K. 1981. Cellular and dynamic aspects of oocyte growth in teleost. *American Zoologist* 21: 325-343.

Weatherley, A.H. and Gill, H.S. 1987. *The Biology of Fish Growth*. London: Academic Press.

Welcomme, R.L. 1967. Observations on the biology of the introduced species of Tilapia in Lake Victoria. *Revue Zoologie et de Botaniques Africaines* L. 26(3-4): 249-279.

Werder, U. and Soares, G.M. 1985. Age determination by sclerite numbers, and scale variations in six species from the central Amazon (Osteichthyes, Characoidei). *Animal Research and Development* 21: 23-46.

West, G. 1990. Methods of assessing ovarian development in fishes: A review. *Australian Journal of Marine and Freshwater Research* 41: 199-222.

Windell, J.T. 1966. Rate of digestion in the bluegill sunfish. Invest. *Indiana Lakes Streams* 7: 185-214.

Windell, J.T. 1971. *Food Analysis and Rate of Digestion*. Oxford: Blackwell.

Windell, J.T. 1978. Estimating food consumption rates of fish populations. *In:* Bagenal, T. (ed.). *Methods for Assessment of Fish Production in Fresh Waters*, pp. 227-254. London: Blackwell Scientific Publications.

Windell, J.T., Foltz, J.W. and Sarokon, J.A. 1978. Effects of fish size, temperature and amount fed on nutrient digestibility of a pelleted diet by Rainbow trout, *Salmo gairdneri*. *Transactions of the American Fisheries Society* 107: 613-616.

Windell, J.T., Norris, J.F., Kitchell, J.F. and Norris, J.S. 1969. Digestive response of rainbow trout, *Salmo gairdneri*, to pellet diets. *Journal of the Fisheries Research Board of Canada* 26: 1801-1812.

Winship, A.J., Trites, A.W. and Rosen, D.A.S. 2002. A bioenergetic model for estimating the food requirements of Stellar sea lions *Eumetopias jubatus* in Alaska, USA. *Marine Ecology Progress Series* 229: 291-312.

Winstanley, R.H. 1990. A fisheries manager's application of minimum legal lengths. *In:* Honcock, D.A. (ed.). *Legal Sizes and their Uses in Fisheries Management*. Australian Society for Fish Biology Workshop, Lorne, 24 August 1990. Bureau of Rural Research Proceedings No. 13. Australian Government Publishing Service, Canberra.

Wöhl, S. and Schuster, S. 2006. Hunting archer fish match their take-off speed to distance from the future point of catch. *The Journal of Experimental Biology* 209: 141-151.

Wootton, R.J. 1990. *Ecology of Teleost Fishes.* New York: Chapman and Hall.

Wootton, R.J. 1998. *Ecology of Teleost Fishes.* The Netherlands: Kluwer Academic Publishers.

Xin, Li., Xiaoyun, He., Yunbo, Luo, Guoying, Xiao, Xianbin, Jiang and Kunlun, Huang. 2008. Comparative analysis of nutritional composition between herbicide-tolerant rice with bar gene and its non-transgenic counterpart. *Journal of Food Composition and Analysis* 21: 535-539.

Yamamoto, K. and Yamazaki, F. 1961. Rhythm of development in the oocyte of the gold-fish, *Carassius auratus. Bulletin of the Faculty of Fisheries Hokkaido University* 12: 93-114.

Yön, N.D.K., Aytekin, Y. and Yüce, R. 2008. Ovary maturation stages and histological investigation of ovary of the zebrafish (*Danio rerio*). *Brazilian Archives of Biology and Technology* 51: 513-522.

Youson, J.H., Holmes, J.A., Guchardi, J.A., Seelye, J.G., Beaver, R.E., Gersmehl, J.E., Sower, S.A. and Beamish, F.W.H. 1993. Importance of condition factor and the influence of water temperature and photoperiod on metamorphosis of sea lamprey, *Petromyzon marinus. Canadian Journal of Fisheries and Aquatic Sciences* 50: 2448-2456.

Zeitoun, I.H., Ullrey, D.E., Magee, W.T., Gill, J.L. and Bergen, W.G. 1976. Quantifying nutrient requirements of fish. *Journal of the Fisheries Research Board of Canada* 33: 167-172.

Zorica, B., Sinovčić, G. and Pallaoro, A. 2005. Age, growth and mortaligy of painted comber, *Serranus scriba* (Linnaeus, 1758), in the Trogir Bay area (eastern mid-Adriatic). *Journal of Applied Ichthyology* 21: 433-436.

Index

For Product Safety Concerns and Information please contact our EU
representative GPSR@taylorandfrancis.com
Taylor & Francis Verlag GmbH, Kaufingerstraße 24, 80331 München, Germany

www.ingramcontent.com/pod-product-compliance
Lightning Source LLC
Chambersburg PA
CBHW070730220326
41598CB00024BA/3378

9 781032 738161